Spacegirl

II

Spacegirl II: 21 women write about their careers on Earth in the Space Industry
Compiled by: Gary L. Gilbert

Published by Flying Goddess Publishing
Florida, USA

flyinggoddesspublishing.com

Contact publisher for bulk orders and permission requests.

Cover & book design by Ghervann Michael Yntig and Leesa Ellis of 3 ferns ➤ 3ferns.com
Cover & book formatting by Leesa Ellis of 3 ferns ➤ 3ferns.com
Self-publishing consultancy by 3 ferns ➤ 3ferns.com

Printed in the USA.

ISBN (Hardcover): 979-8-9899700-6-3
ISBN (Paperback): 979-8-9899700-7-0
ISBN (eBook): 979-8-9899700-4-9

Spacegirl

II

21 WOMEN WRITE ABOUT THEIR CAREERS ON EARTH

IN THE

SPACE INDUSTRY

COMPILED BY Gary L. Gilbert

Flying Goddess

PUBLISHING

That's one small step for man,
one giant leap for mankind.

– NEIL ARMSTRONG

Contents

Spacegirl

PROLOGUE

Four hours, 55 minutes, and 23 seconds – the time John Glenn spent orbiting the Earth alone in the Mercury spacecraft he named *Friendship 7*. With the help of numerous people, in a ground-breaking historical journey on the 23rd of February 1962, he became the first American to orbit the Earth three times. Could it be connected to the fact that I was one year old on the day John Glenn embarked on that miraculous voyage? Or it could be the fascination I developed growing up while watching Star Trek movies and the gripping television series, "Space: 1999". Whatever it was, it had me hooked! The flames of my ongoing interest in everything related to space were ignited, fueling the dreams that have accompanied me into the future.

I have followed my passion or obsession with space travel throughout my life. I am convinced that the ongoing existence of the human race depends on our future being comparable to that portrayed in "Star Trek: The Next Generation" and the different shows that came later. During 2011 – 2012, I attended the International Space University and was asked to give the graduation speech. Looking at the sea of faces and their sensational achievements, I was drawn to something that made a distinct impression. What stood out was that from the class of 32 or more people, 10 or 11 of those graduates were women.

That ceremony prompted me to propose and suggest that one of the incoming or graduating class members might be stepping onto Mars or be the first female NASA Administrator.

After graduation and returning home, I remember watching an episode of "Star Trek: The Next Generation" where Geordi La Forge, the chief engineer of the USS Enterprise-D, was talking on the halo deck with an engineer called Dr. Leah Brahms. It occurred to me that this female character played an extremely major part in ensuring the capability of the warp drive that propels all of the ships of the federation vessels.

As a female human scientist of the United Federation of Planets, she contributed significantly to developing the Galaxy-class's warp drive systems. This episode impacted my thinking about our real-life space program and the roles women are involved in. It was then I decided there was the potential and perfect opportunity for more females to take lead roles in the space program, not only as astronauts but in the crucial role of engineers.

I looked at the careers and lives of inspirational Janet E. Petro, the First female Director of the John F Kennedy Space Center; Charlie Blackwell-Thompson, the First Female launch Director at the Kennedy Space Center; and Breanne Stichler, one of the six people currently on this planet to operate the mobile launcher.

What makes these women so remarkable and influential? They contribute to an integral part of the mission that will take astronauts back to the moon, Mars, and beyond.

Each of these women has an indispensable role in the voyages that will take place, and they hold key positions in the space program. While they are not the ones who will be the first to set foot on another planet, without their knowledge, expertise, and commitment to the space program, success for future missions may not be realized. Behind every astronaut like John Glenn, there have been women behind the scenes who are owed more recognition for what they have achieved and the success of current and future missions. The future looks exciting, and possibilities abound!

There are plenty of excellent books about astronauts – mainly with men in the lead role. They often act as inspiration for children and young people who are interested in becoming an astronaut. I decided to write Spacegirl to shine the light on various women in the space program, and it is a collection of their stories, achievements, and lives. The hope is that it will inspire eighth- to twelfth-grade girls to look further into the space program at other crucial opportunities and possibilities, rather than with a sole focus on becoming astronauts.

"I think the most rewarding thing about what I have done is that it opens people's minds to the possibilities of what they can do. "

— JOHN GLENN

Laia Lopez

"There are no limits for the ones that dare look beyond."

– LAIA LOPEZ

Have you ever wondered why we exist? How big is the universe? And how did we end up being part of it? When I was a kid, I dreamed about finding answers to these questions. I loved science subjects and mathematics even though I was told that girls were good at humanities. I felt a bit different than my classmates, but I never let that stop me. I was proud of liking science.

It was great to grow up with teachers that supported me. I still remember how me, and some friends would stay after class to discuss questions and curiosities we had with the teachers. Of course, there have also been times where I have received bad comments from other classmates. It hurt me a lot. Especially when you are a teenager, and you are simply figuring out who you want to be.

My school incentivized students to join math and science competitions, some of which we won. It was rewarding to see that all that effort paid off but, sometimes, it was exhausting too. I was really lucky when, at the age of fourteen or fifteen years old, my parents searched for scholarships for me and found one called Youth and Science, a three-year research summer program in which students join research institutions every summer where they learn from experts. After the different rounds of the selection process – letter of motivation, CV, interviews, etc – I was told that I had been selected. That was when my life completely changed, and I could never thank the Youth and Science foundation enough for giving me this opportunity.

Every summer, I went to a different research institution. The first year, I went to MonNatura Pirineus, located in the mountains in the north of Catalonia, where I joined the astronomy group. Our professors were PhD and Postdoc researchers from the University of Barcelona. We were a class of around ten students that spent those summer nights observing galaxies, star clusters and detecting exoplanets. Those nights sitting in the fields with a couple of telescopes, a blanket and the best company were nights I will never forget.

During the second and third year of the program, I went to the International Astronomical Youth Camp, the astrocamp, and I also joined the Route of Stars in the Teide Observatory of the Canary Islands. During those three summers, I met like-minded students that shared the passion for astronomy and space.

At the end of high school, I had to deliver a research project. I wanted to do something original and challenging. My project was called "Can I Calculate the Speed of Light?" It consisted of calculating the speed of light by doing astronomical observations of Jupiter's satellites period and calculating the speed of light via Ole Rømer's method. It was also a good excuse to get my parents to buy me a small telescope.

After finishing this project, the most important time for a teenager came. That is the time where you need to make a choice of what's next. You need to decide what career you want to pursue. Since I loved math, my first obvious answer was that I wanted to become a mathematician. Then I realized that I actually wanted to study something that would allow me to answer all the questions that were always in my head: why do humans exist? How big is the universe? How was it created?

I thought of studying philosophy, but then I realized that physics is actually the perfect mix between mathematics and philosophy. I want to take this opportunity to share that at that time, I felt that it was a decision that only had one correct answer. Later, I realized through my family and friends' experience that it is okay to change careers if you realize that your objectives have changed.

In September 2016 at the age of seventeen, I entered through the door of the Faculty of Physics of the University of Barcelona. My university period was definitely the most challenging period of my life. I chose to study the theoretical branch of physics, in which we studied astrophysics and cosmology (among many other disciplines). We started the degree with two hundred students and ended with only seventy.

During my free time at university, I used to give classes to middle school and high school students with difficulties, and during the summers, I would volunteer at the observatory of my city. The observatory was right in the middle of the city, and as you know, there is a lot of light pollution in cities, so it wasn't the best place for doing observations. However, I did guided tours, observation sessions for families and workshops for kids about the Solar System, black holes, the Moon, and supernovas, etc. Honestly, those moments were the best. There is no better reward than seeing the eyes of the little kids dreaming about stars, planets and truly enjoying your explanations when you talk about the universe. It filled me with so much joy.

After finishing my physics degree, I wanted to do a master's in astronomy or something related to the space sector, but I didn't really know exactly which master I wanted to pursue or where. One thing I realized is that when you study, they don't usually teach you how the industry works. Universities mainly focus on teaching you the theory. You don't usually learn what are the main companies in the industry, how it is structured, or how the job environment is, etc.

That's when I found the International Space University (ISU) and its master of Space Studies. ISU fosters an international, interdisciplinary, and intercultural environment and it shows its students the space industry inner workings. It was amazing to find a place full of professionals and students from all around the world that share passion for space. It was during my masters at ISU in France that I learned a lot about the International Space Station.

Yes, we have a space station that is flying above our head 400 kilometers high. Astronauts have been living in the space station since 2000. Of course, they are not always the same ones. Each group of astronauts spends approximately six months onboard the space station. This amazing project is a collaboration of many different countries and space agencies such as NASA, the Canadian Space Agency (CSA), the European Space Agency (ESA), the Japanese Space Agency (JAXA) and the Russian Space Agency (Roscosmos). As one of my professors said, the space station should receive the Nobel Peace Prize.

I am European so you can imagine how happy I was when I learned that the control center that performs operations of the European module (the Columbus module) was in Munich, Germany. After hearing this, all I could think of was finding a way to get an internship there. It was quite difficult to achieve as the COVID pandemic was still very present at that moment. After talking to different people involved in the project, I managed to achieve it. The internship was mostly online but I was lucky to visit the Columbus Control Center one week.

After finishing my internship, I applied for a job at the facility and well, now it's been almost two years working for the Columbus Control Center as a ground controller. My daily tasks consist of monitoring and controlling all the ground systems used to perform space operations in the Columbus module. Our team works 24h 7 days a week. To do that, we divide ourselves into three shifts: morning, afternoon, and night shift. Sometimes it's a bit tiring to be working during the night, but being able to listen to the astronauts and see their experiments and the research performed on board is worth it. I also had the opportunity to meet a couple of astronauts in person. I would have never thought that was a possibility or that I could pursue a career like this. I would not change for anything every single step that took me here. I don't know where life will lead me, but I know I will be close to the space industry.

I want to send a message to all the small girls reading this. You should pursue the things you like. If you want, you can pursue a career in science and technology. Just make sure that what you choose is something you really enjoy, and if it is not, you can always change the direction of your life. I hope my story can help you lift off your career.

There are no limits for the ones that dare look beyond.

Filipa Barros

66 The universe eagerly awaits all those who dare to seek its secrets. 99

– FILIPA BARROS

My first encounter with the wonders of space happened when I received a tiny, toy-like telescope from my dad. The moment he presented it to me, my young heart filled with excitement. Little did I know that this seemingly insignificant gift would ignite a profound curiosity within me, setting me on a lifelong path of exploration and discovery. As I grew older, my fascination with space only intensified. Throughout high school, I dedicated myself to studying technology and science, immersing myself in subjects that would lay the foundation for my future endeavors.

While my passion for space remained unwavering, I also pursued diverse interests such as horseback riding and debating, realizing that we don't have to be defined by a single pursuit. With an insatiable thirst for understanding the universe, I eagerly absorbed knowledge about the cosmos, astronomy, and electronics. This thirst eventually led me to pursue a degree in electronic engineering with a focus on satellite communications. During my studies, I had the opportunity to interact with like-minded individuals, especially during a year abroad where I connected with people who shared my passion for space. This exposure to different perspectives further fueled my enthusiasm for exploration.

Now, as a PhD student in computer science applied to astronomy, I find myself at the captivating intersection of two fields. Through my studies, I have discovered a whole new dimension of exploration, where I can combine my programming skills with my love for the cosmos. As I delve deeper into my research, I am constantly reminded of the vastness of space and the infinite mysteries that await unraveling. Our current understanding of the universe merely scratches the surface, and there is an entire cosmos waiting to be discovered. In addition to my academic pursuits, I am also fortunate to be an invited coding lecturer at the university where I study. This role allows me to share my knowledge and passion with aspiring students, igniting the spark of curiosity in their eyes.

Witnessing their growth and knowing that I have played a small part in shaping their own journeys of exploration brings me immense joy. However, my journey in the field of space science extends beyond the confines of classrooms and laboratories. I have been blessed with extraordinary opportunities that have taken me to unexpected places and exposed me to unique experiences.

One such experience involved participating in a simulated diving astronaut mission, where I had the privilege of simulating the challenging conditions astronauts face during extravehicular activities. This hands-on adventure not only deepened my understanding of space exploration but also instilled a sense of adventure within me, fueling my desire to push the boundaries of human knowledge even further.

Furthermore, my path has been enriched by the guidance and mentorship of seasoned astronauts during a summer school focused on space exploration. Interacting with these individuals who have ventured into the great beyond has been both awe-inspiring and surreal. Their wisdom, combined with their firsthand accounts of space travel, has provided me with invaluable insights into the challenges and wonders of the cosmos. Their stories have solidified my determination to contribute meaningfully to the field and make my own mark in the realm of space science.

Currently, I am engaged in an exciting research project centered around predicting solar activity. The sun's behavior has the potential to disrupt our communication systems and pose risks to astronauts, making understanding it of paramount importance. Leveraging my skills in computer science, astronomy, and data analysis, I am diligently developing models that can accurately forecast periods of heightened solar activity. This research not only serves to safeguard our technological infrastructure but also contributes to the ever-growing body of knowledge about our closest star. Additionally, my research allows me to travel extensively, providing opportunities to meet people from diverse backgrounds, further enriching my experience.

Reflecting on my journey in the field of space science, I am filled with gratitude for the remarkable experiences I have had. Each step, from that small toy telescope to my current research, has brought me closer to unraveling the mysteries of the universe and pushing the boundaries of human knowledge. I am humbled by the opportunities that have come my way, and I eagerly anticipate what the future holds as I continue to explore the vast expanse of space – both through the lens of scientific inquiry and my own unquenchable curiosity. With each discovery, I aspire to inspire others to gaze up at the night sky with wonder and embark on their own journeys of exploration. The universe eagerly awaits all those who dare to seek its secrets.

Amy Kaminski

66 Stay flexible and be open to evolving and exploring your interests, and you will find a fulfilling path for yourself. 99

– AMY KAMINSKI

My professional journey has been somewhat unique. I am not a practicing scientist or engineer. Instead, I have spent part of my career creating policies and making recommendations to government decision makers about what the United States should do in space and how we should go about it. More recently, I have focused on developing ways to involve everyday people in helping NASA with its scientific research and technology needs to explore space. I have enjoyed my work in these areas because they have aligned with my skills and interests – including my love of space.

I started getting excited about space when I was eight years old. One August evening, my grandfather invited me to watch the Perseid meteor shower in my grandparents' yard. It was the most amazing show I had ever seen. Although it was quiet and slow, I was mesmerized by each streak of light across the dark Pennsylvania sky. Around that time, I was also spending a lot of time reading about science. My parents had a collection of little books on topics like trees, insects, and dinosaurs. One was about stars, and I remember spending hours poring over that particular book.

Around third grade, something pivotal happened: I was at a Girl Scout meeting and a friend told me there was a camp kids could go to and learn all about space. This was in the days before the internet so my parents had to do a little research, but they got hold of a brochure for the US Space Camp in Huntsville, Alabama. It was not an inexpensive camp, so I felt very fortunate when my parents allowed me to attend when I was eleven. Trying out the space flight simulators, building model rockets, and seeing movies about space shuttle missions felt to me like a week in paradise, and I returned home with the dream of becoming an astronaut one day.

I spent my middle school and high school years doing all I could to prepare for my future in space. I continued to do as much reading as I could on astronomy and space exploration and I decorated my bedroom walls with posters on space shuttles and astronaut mission crews. But most importantly, I was very lucky to find a special mentor – my seventh-grade teacher, Mrs. Palazzolo – who inspired me to keep pursuing my dream. Mrs. Palazzolo loved space – in fact, in the mid-1980s, when NASA had invited teachers to apply to fly on the space shuttle, she had been selected as one of about 100 finalists for the flight. Although she did not get to go on a mission, she shared her passion by teaching a space exploration class, which I absolutely loved. During high school, I further fueled my interest and knowledge by volunteering at our city's planetarium and science center. I also participated in space engineering and astronomy programs for high school students at two universities before my senior year.

College was the time where I really learned a lot about my specific interests in a space career. I decided to go to Cornell University because it was known for its space science department (the famous astronomer Carl Sagan was still on the faculty when I enrolled). I also liked that Cornell had a major field of study I had not seen at many other schools called science and technology studies. It allowed students to explore how human decisions shape science technology and how, in turn, science and technology acted as powerful forces in society. I took classes in the history of science, science journalism, and science policy.

After several semesters, I found that I enjoyed these classes even more than the math, physics, astronomy, and geology classes I was taking! I started to think about how I could mingle these interests in a space-related career. I came to the decision during my senior year of college that my heart was not in a career in scientific research or engineering – the two career paths other than medicine or military service that at the time qualified one to apply to be an astronaut. So I made peace with the fact that I would revector my dream of becoming an astronaut into some other type of space career.

During the summer before my senior year, I had participated in a NASA internship program where I learned a lot about how decisions about the space program get made by government leaders in Washington, DC. I also heard that The George Washington University offered a master's degree program in space policy, where one could train to become part of the community of people who help make and advise on the direction of space activities. I decided enrolling in the program would be a good next step in helping me blend my interest in space with the social sciences.

There are many different actors involved in the space policy community across the federal government as well as in companies, nonprofit organizations, and universities. When I graduated, I knew I wanted to be in the middle of the action and ideally at NASA. The problem was that NASA had paused hiring temporarily at the time I was looking for a job. I applied for positions with other government agencies, but getting hired by the government can take a very long time. So, I joined an organization called the National Space Society, which works on behalf of members of the public who are enthusiastic about space to convince Congress to support space activities. It was fun to deliver stacks of petitions signed by the Society's members to the offices of key members of Congress.

After four months, however, I got hired for my first federal position – as an analyst at the Federal Aviation Administration. I spent two years there, developing reports on vehicles that had been developed by the U.S. aerospace industry and the satellites they might carry. Then an opportunity came up that would get me closer to NASA: the White House Office of Management and Budget (OMB) was looking for new people to join the team that determined what NASA programs and budgets the President of the United States would propose each year to Congress. It was tremendously exciting to work there because I was able to work with many talented people and directly impact projects ranging from the Hubble Space Telescope to satellites that are monitoring climate change on Earth today.

I loved my job at OMB but left after eight years when a couple of things changed. One was that I had the chance to take a special assignment at NASA in the office that developed rovers and other spacecraft to explore the solar system. At long last, I had a pathway to working at NASA! Also, around this time, I decided to return to school to pursue a Ph.D. I wanted to deepen my knowledge in the field of science and technology studies, and I found a program at Virginia Tech that would allow me to take classes while continuing to work. It was not easy, and eventually I reduced my hours at NASA, but after five years I completed my degree. As part of a Ph.D. program, one must conduct a major research project. I studied why and how NASA involved different types of people in the space shuttle program.

That project, and the classes I took, helped me hone a passion in how the public relates to space activities. At NASA, I had since moved to become an advisor to the chief scientist and I started to examine how the space agency was involving the public in research through a range of "citizen-science" projects. A few years later, I moved into a role to develop contests for the public to propose solutions for NASA technology needs. Now I am leading a team of talented professionals to share information about NASA's scientific findings and to invite the public to participate in making future discoveries.

I have thoroughly appreciated being able to put my academic education and interests into practice at NASA in initiatives I find very meaningful. I have also enjoyed authoring several publications about NASA's public engagement work to help others understand what the space agency has achieved. One day I hope to teach college and graduate courses in this area.

My experience illustrates that many types of space-related careers are possible, and there is no single "right" path to pursue. It is important to keep thinking about how to achieve your dream, but it is ok if that dream changes along the way! It is natural that your interests may change, or you may want to combine multiple interests. Stay flexible and be open to evolving and exploring your interests, and you will find a fulfilling path for yourself.

Anne Marie Robinson

"I feel I am at the dawn of a new era in my country of space exploration and hope to join the ride."

– ANNE MARIE ROBINSON

A long time ago in a country far, far away, a crazy Kiwi Kid started her journey into space. The journey took her on amazing adventures all over the world. This is her story.

Growing up, I asked my parents for a telescope but had to make do with binoculars so I could look at the stars from my second-story bedroom. From early on, I knew I wanted to go into space, even to the Moon. Voicing this career choice got many a laugh and scoff over the years but it didn't change my plans.

At school I took as many science subjects as I could, gained my pilot's licence when I finished school, went to university to gain a science degree, and then worked in the aviation industry as a flight attendant, and then with the Royal New Zealand Air Force Museum to promote and immerse myself in aviation.

It all began in 1979 when my dad went to Antarctica for a six-month deployment as a deputy officer in charge of Scott Base, the New Zealand science station. He met NASA scientists looking for meteorites in the snow and got a contact address for me, which he sent home. In those days, mail was the favoured form of sending information, even from Antarctica.

I wrote to NASA, asking about becoming an astronaut. I was sent information and an application form, which I filled out and sent back. Of course I was rejected as you needed to be a United States citizen to work in the NASA space programme. However that did not deter me, so I applied once more over the years and was again rejected.

The NASA public affairs officer who I'd initially corresponded with became a very good friend and, over the years, I made many trips to the United States to meet him and stayed with his family in Houston Texas. He became like my second dad.

He sent me a brochure about the United States Space Camps in Huntsville Alabama. These were simulated astronaut and fighter pilot training camps. I was instantly hooked and applied to attend my first camp in 1989.

After being accepted, and then sourcing sponsorship and funding from around New Zealand, my first trip to the United States was to visit the Johnson Space Centre in Houston Texas where my liaison public affairs officer was based. He toured me around the Space Centre, introduced me to astronauts and allowed me to work with mission controllers on simulated space missions. I remember putting on the first type of VR headset and a glove to experience virtual reality NASA was developing. I put my hand into an astronaut glove and then into a vacuum box to experience how difficult it was to use fingers in the bulky space suit in a vacuum.

I then continued on to space camp, attending with thirty other trainees. I was the only New Zealander. I was space shuttle commander for one of the space camp missions. I joked with my crew that a New Zealander in charge of an American space shuttle was ironic after applying to NASA to be an Astronaut and being turned down because I was not a US citizen. Other positions I trained in were mission specialist on a space station and a payload specialist on another space shuttle mission, where I got to do a simulated EVA (extra vehicular activity) and swing over the cargo bay that was held by cables from the ceiling. At one of the Space Camps, I was riding the MMU (Manned Manoeuvring Unit) when the machine malfunctioned and I was left dangling upside down while staff found a maintenance crew to repair the machine. It took a while before the machine could be repaired and I was freed but that's space for ya! I think the counsellor was more concerned than I was but all in a day's work, as they say.

When I returned to New Zealand, the news media had found out about my endeavours and started asking for interviews, public appearances, lectures, talks, and wrote newspaper and magazine articles about the Kiwi Kid trying to be an astronaut. This started a journey of many years talking and promoting the United States space programme all over New Zealand. There is natural human awe about space so it was a subject that both excited and attracted huge interest that I was able to support.

It is very expensive to travel anywhere in the world from New Zealand so I was always looking for support and fundraising to attend Space Camp. One trip, I managed to get sponsorship from one of the American Airlines that flew to New Zealand. They even upgraded me to first class on the way there which was incredible to experience. I was also very privileged to be allowed the jump seat on the flight deck as we flew into Los Angeles Airport. The aircraft was the old style 747 so sitting in the flight deck meant you were about two stories high when the aircraft landed so I was still looking down at the runway in awe of this amazing aircraft and enjoying the ride.

Over the years, I returned to NASA almost every year for over twenty-five years – the last trip being in 2016 where I attended space camp with my daughter. I was truly privileged to be hosted by NASA on many trips to the Johnson Space Centre, meeting astronauts, flying the space shuttle simulator, and attending space camps.

Due to the media attention, I was hired as a volunteer United States space camp ambassador for New Zealand in 1991. Space camp has ambassadors from many countries around the world who promote and encourage people to go to space camp. As an ambassador, I would travel around New Zealand talking about the US space programme and promoting space camp. I had a group of enthusiastic people who had a common interest in space exploration and aviation that came with me. We were sponsored to various events and locations where we would set up displays and talk with people. I called these amazing friends the New Zealand Space Camp Team, all volunteers and all very dedicated to promoting Space exploration.

On another trip to space camp, one of the camp counsellors offered to drive me to Florida to watch a space shuttle launch. It was a nighttime launch and we were positioned several miles away among many hundreds of people, sitting and standing on the grass area with everglades in front of us. When the space shuttle launched, the noise was amazing but the light from the engines lit the sky as though it was daytime. I will never forget it.

In 1994, I negotiated with NASA to bring space exhibits and an astronaut to New Zealand to work with a science centre in Christchurch and tour around the country. This was a very successful event with many hundreds of people coming to meet the astronaut and enjoy the exhibits. One of the exhibits was a mock-up of the MMU (Manned Manoeuvring Unit) and the space camp team would host groups of people and explain how the equipment worked in space. I would coincidentally meet this astronaut again many years later when I was at Space Centre Houston in 2016, attending an astronaut talk and he was the astronaut.

Several years later, the NASA public affairs officer arranged for me to fly on NASA SOFIA when it was in Christchurch. SOFIA would come to Christchurch every July to explore space through the clear skies over New Zealand. The flight was amazing. External aircraft doors would slide open to expose the telescope and I watched over the shoulders of the scientists as they targeted the areas of space they wanted to look at and the computers recorded the data.

In 2000, my husband and I received an invitation from the NASA astronaut I had brought out to New Zealand to watch his space shuttle launch from the Kennedy Space Centre.

We managed to scrape together air tickets and accommodation for such an exciting opportunity. We were treated as VIP guests when we arrived at the Kennedy Space Centre and met his wife, who hosted us for dinner while her husband was in quarantine prior to launch. Disappointingly, the launch was delayed due to a tool being left on the launch platform and it was rescheduled several days later. We couldn't wait due to our itinerary so we ended up watching it on television in Texas.

My daughter was born in 2006 and I had to change my employment so I could be a mum but still promote and participate in space adventures. I went back to university and trained as a school teacher and designed an outreach programme called Kids in Space. Schools would book me to come and teach students about space. Over the years, I built up a collection of astronaut flight suits, videos, and STEM interactive activities that students could participate in. The joy of watching excited students dress up as astronauts and learn about space exploration was inspiring.

In 2013, the NASA SOFIA aircraft was back in Christchurch conducting telescope observational studies. NASA contacted me to invite us to tour the aircraft while it was in town. I took my family and my dad on the tour. As we boarded the aircraft, we were met by NASA staff and introduced to the scientists working on computers and planning the next telescope viewing flight. Suddenly one of the staff called us over to the middle of the aircraft and admitted it was a bit of a ruse asking me on the aircraft. They presented me with a plaque and NASA medal (an exceptional public achievement medal) as the scientists gathered around and applauded. All the years I had spent promoting NASA in New Zealand had not gone unnoticed and I suspected my friend, the public affairs officer at the Johnson Space Centre, had something to do with this. I was so shocked and so proud, and to my knowledge I am the only New Zealander to receive a NASA Medal. As my dad had started my space journey, it was really cool he was there to see this awesome moment. I take the medal to show students at schools.

Through my work with the Royal New Zealand Air Force Museum, the New Zealand Space Camp Team attended and worked at several air shows called Warbirds over Wanaka in the South Island of New Zealand.. These bi-annual air shows attracted vintage and modern aircraft from all over the world to display and impress thousands of people in air combat and aerobatic skills. At one of these air shows, I was able to negotiate to host Apollo 11 astronaut Buzz Aldrin as a special guest. We had the privilege of looking after Buzz and his wife while he was at the air show. One of our team even managed to get Buzz to leave an imprint of his boot on a piece of paper, which she still has to this day.

All these years later, my space adventure continues. I have met and made many friends in the American Space Industry who took the time to support and educate me about their Space Program. I am still in touch with many of them. Due to their kindness, it has made me who I am and helped support my space endeavours all these years. Per aspera ad astra (through adversity to the stars)

While I have never gone into space, the journey to get close has been exciting and I am privileged to meet amazing people and to learn and experience wondrous things. New Zealand is at the end of the world but is slowly making its own way in the Space Industry with Rocket Lab, Dawn Aerospace, and Kea Aerospace. I feel I am at the dawn of a new era in my country of space exploration and hope to join the ride.

Shelli Brunswick

"One of the most exciting aspects of this... is creating pathways for the next generation of global thought leaders."

— SHELLI BRUNSWICK

Embracing the Unknown: Launching S.B. Global LLC

In 2024, I took a significant leap by founding S.B. Global LLC, an extension of my passion for space technology and my commitment to creating access for all. S.B. Global LLC isn't just a business venture – it represents the culmination of decades of experience in the aerospace and space industries, from my early days in the U.S. Air Force to my leadership at Space Foundation. The company embodies my vision to bridge the highly specialized world of space and the general public, making space technology more accessible, understandable, and, most importantly, inclusive.

As the Founder and CEO, my focus has expanded from education and collaboration to championing underrepresented groups and fostering new opportunities within the global space ecosystem. I engage with governments, businesses, and educational institutions worldwide to promote space technology innovation and ensure its benefits are available to everyone, regardless of background or circumstances.

One of the most exciting aspects of this chapter of my career is the opportunity to pursue a deeply personal mission: creating pathways for the next generation of global thought leaders. Space has become an integral part of life on Earth, touching everything from healthcare and communication to environmental sustainability and disaster management. I've witnessed firsthand how space can transform lives, open doors, and solve some of the world's most pressing challenges. I'm dedicated to ensuring that the future of space and its adjacent industries are built on a foundation of inclusivity, collaboration, and shared innovation.

My goal is to ensure that opportunities and advancements in space technology are accessible to all people and that the space industry reflects the diversity and richness of the global population. This means creating more opportunities for women, people from underrepresented communities, and young professionals to enter and thrive in the space industry.

Becoming a Global Advocate: Connecting Space to the Public

One of the most rewarding aspects of my journey has been my role as a global advocate for the space ecosystem. Space exploration is often viewed as the domain of scientists, engineers, and astronauts. However, the innovations that emerge from space – whether in medicine, agriculture, climate science, or communications – profoundly impact every person on Earth. Yet, space still feels distant and unattainable for many people, reserved for those with specialized knowledge or access to resources.

My mission is to change that narrative. I focus on connecting the intricate world of space technology with the broader public, making space feel tangible and relevant.

By acting as a bridge between the specialized knowledge of the space sector and the everyday realities of people's lives, I help foster a greater understanding of how space innovation can be used as a tool to solve some of the world's biggest challenges. From providing satellite data to monitor climate change to developing new medical technologies, space has the potential to benefit everyone.

This advocacy work is grounded in my belief that the space industry must be inclusive to thrive. The more diverse voices we bring into the conversation, the richer our ideas, solutions, and innovations will be. I've worked to form partnerships that create opportunities for underrepresented groups – particularly women and young people – to enter the space and technology sector.

The challenges we face in the space industry are vast, from developing sustainable space travel to understanding the impact of space debris. These challenges, like many others, require the creativity and input of people from all backgrounds. By encouraging more people to explore careers in space, we can ensure that the industry remains innovative, dynamic, and prepared to meet these and future challenges.

The Power of the Stage: Speaking to Inspire and Empower

Throughout my career, the power of storytelling has remained constant. As a futurist and motivational speaker, I've shared my journey, insights, and passion for space technology with audiences worldwide. From global conferences to corporate boardrooms, I've spoken about the boundless opportunities within the space sector and the potential that space holds for solving some of the world's most complex problems.

Throughout my speaking engagements, I've realized that my message resonates deeply with people from all walks of life. Space, after all, is not just about launching rockets or exploring distant planets – it's about using the technology and lessons we develop through space exploration to improve life here on Earth.

Whether I'm discussing the role of satellites in monitoring climate change or the medical advancements derived from space research, I always emphasize how interconnected space and Earth truly are.

One of the key themes in my talks is the importance of perseverance – seizing opportunities even when the odds are stacked against you. My entry into the space industry wasn't planned, and I didn't grow up dreaming of working in this field. I initially resisted my Air Force assignment to space acquisition, but this unexpected opportunity began a career I never imagined – one filled with opportunities to lead, innovate, and advocate for change. Through my speeches, I share these personal experiences to encourage others to take risks, embrace the unknown, and trust that every step – whether planned or not – can lead to remarkable new horizons.

I also emphasize the importance of collaboration and inclusivity. The space industry is vast, multifaceted, and complex, requiring people from all backgrounds to come together to solve problems. My goal is to inspire people to see space as an exciting frontier and a place where they can make a meaningful impact on the world.

Bridging Worlds:
Advocacy and International Collaboration

In addition to my work as a speaker, I've focused on building connections across diverse sectors through my advocacy and brand ambassador roles. By its very nature, space is a global endeavor, requiring collaboration across borders, industries, and disciplines to truly thrive. The future of space exploration will depend on our collective ability to work together, leveraging the strengths of each sector to achieve shared goals.

One of my proudest achievements has been working with international organizations such as the UNOOSA Space4Women Mentoring Program, the WomenTech Network, and the World Angeles Investment Forum Global Women Leaders Committee. These organizations share my commitment to fostering opportunities for women and underrepresented groups in space technology.

Together, we've been able to create programs, speaking engagements, and mentoring opportunities that unlock innovation and bring fresh perspectives to the table.

Through advocacy, I've seen firsthand how global collaboration can open doors and lead to new innovations that benefit the space industry and the world. Whether through mentoring programs, international partnerships, or collaborative research projects, my goal is to ensure that the future of space exploration remains accessible and beneficial to all. By working together across borders and industries, we can unlock space's full potential to solve some of the world's biggest challenges.

What's Space Got To Do With It? My Book Series

As part of my mission to make space more accessible to a broader audience, I'm thrilled to introduce my new book series, *What's Space Got To Do With It?* The first book in the series, *10 Life Lessons for Personal Growth,* explores how space technologies and principles can inspire personal and professional development.

Each book in the series will include insights from global thought leaders I've had the privilege of meeting and interviewing throughout my career. These individuals represent diverse industries, regions, and backgrounds, from space exploration to business, policy, finance, and beyond. The lessons these visionaries share offer a rich perspective on leadership and innovation grounded in real-world experiences.

Future titles in the series, such as *An Interstellar Guide to Success and Out Of This World Leadership,* will explore the strategies and principles that have guided my career and those of other global leaders across industries. These books are not just for those working in space – they are for anyone seeking to push the boundaries of what's possible in their lives, whether in business, science, or personal growth.

I hope that readers will see through this series that the lessons we learn from space exploration – perseverance, collaboration, and innovation – are applicable across all fields and challenges. Space is a metaphor for what we can achieve when we dare to dream big and take risks. Whether you're launching a business, leading a team, or overcoming personal challenges, the principles that have guided space exploration can inspire you to reach new heights.

Looking to the Future:
Building a More Inclusive Space Ecosystem

As I look ahead to the future of the space ecosystem, I'm filled with optimism and excitement. The space industry is at a tipping point, with new technologies, new players, and new possibilities emerging daily. But with this growth comes responsibility. It is up to us to ensure that the future of space reflects our global community's diversity, creativity, and resilience.

I remain committed to building that future by continuing to create pathways for underrepresented groups to enter the space industry. I will continue pushing for innovation that serves the world through my advocacy, partnerships, and outreach initiatives. Through my work – whether speaking, writing, or mentoring – I want to help others see the limitless potential of space and the boundless potential within themselves.

The journey I've taken – first as a reluctant participant in the space industry and now as a passionate advocate for its growth – mirrors the journey many others will take in the coming years. The space industry will continue to evolve; with it, there will be opportunities for people from all backgrounds to contribute to its success.

My goal is to inspire others to embrace the vast possibilities of space and see their contributions through my speaking engagements and my writing.

Arlene Gallegos

> **"My aerospace career has had many winding paths and all of them were possible because I was willing to take some risks and get out of my comfort zone."**

— ARLENE GALLEGOS

My path to an aerospace career began when I was 20 years old and working as a cashier at a small retail store called Yellow Front in Pueblo, Colorado. I was enrolled in what was then the University of Southern Colorado completing my prerequisite classes and the time was near for me to choose a major.

I was strongly leaning toward a nursing major because my Aunt Margie was a nurse. I looked up to her and would sometimes see her come home from work in her crisp white nursing uniform. When she would tell me stories about her workday, it seemed to be really exciting and she was helping people! Which to me was the best part of her job. What I didn't realize as a ten-year-old was the very long shifts she worked and, as I look back, I don't remember her telling me any stories about the very stressful and sometimes heartbreaking tasks of being a nurse.

It was during my anatomy and physiology class, specifically working in the lab with a cadaver, that learned I was definitely not cut out for the nursing field. Many times I would have to leave the lab to compose myself because I had a physical reaction to the tasks we needed to do. Sometimes I didn't go back. This was not good! How could I be a nurse if I wasn't able to complete this class? Looking back on this now

I realize it was fortunate that I found this out about myself so early. It forced me to look into other possible careers.

When I was in high school, I thought I would like to go into the Air Force because I loved airplanes. Each year, our family would fly to California for vacation and I would marvel over how the airplane stayed in the air. I didn't understand anything about it, I just thought it was the coolest thing in the world. I wanted to fly them.

Fast forward to high school and, when I told my counselor this, it was recommended to me that I become a secretary or something along those lines because the Air Force didn't allow women to be pilots. Well, hearing this just deflated me. Actually, the Air Force did allow women to become pilots, but the opportunities were few back then. But as the saying goes, everything happens the way it's supposed to happen.

One evening, while watching the nightly news, I saw that a large company was moving into one of the buildings near the airport. It was a company called McDonnell Douglas. I knew that McDonnell Douglas built airplanes, so I assumed that was what was going to be built there. But the work that was moving there was actually the manufacturing of Delta II launch vehicles. I had never heard of the Delta II launch vehicle. Was it like the shuttle? Or was it a missile?

I started researching it and found out that it was a one-use rocket that launched various small-sized satellites. Now this was interesting! So, I decided that I was going to get a job there. I spent a lot of time sprucing up my resume and hand-delivered it to McDonnell Douglas since this was before email and internet.

After many weeks of checking back in person (I was very persistent), I was finally granted an interview and lo and behold, I was offered a job as a manufacturing technician.

McDonnell Douglas provided many weeks of training and then I and the other newly hired technicians began working on the assembly floor. To this day, over 30 years ago, I still remember the joy and excitement I felt, knowing I would be building the Delta rocket!

McDonnell Douglas' policy was to cross-train technicians in all manufacturing and test areas so throughout the years, I worked in several different assembly areas drilling holes, bucking rivets, building assemblies, and building and installing wire harnesses.

Later, McDonnell Douglas also brought MD-80 window work and also commercial aircraft blanket work to Pueblo so I was able to work on those assemblies. I also worked in production control for a short time doing inventory work and filling part orders.

After many years of working as a technician, I moved over to the quality assurance department and have been working in the quality realm ever since. Having all of this manufacturing experience was the best foundation I could have ever had for my quality career.

McDonnell Douglas Pueblo later became Boeing. When Boeing decided to move the Delta work to Huntsville, Alabama, I decided to accept a Quality Assurance Field Inspector position with Boeing at Vandenberg Air Force Base in California.

Moving to California alone was quite out of my comfort zone. I was a small-town girl and never thought I would leave the town I grew up in and loved. My entire family and friends were there. I was scared, but my Dad always taught me to do what I fear. His words were in my mind, "You'll never know what you can do if you don't try."

So off I went to live and work in California. I was then a part of launching the very rockets that I was building years before! It was an exciting six years. Boeing had become United Launch Alliance with Lockheed Martin and when layoffs were happening, I decided to move to Florida to work for Boeing again as an International Space Station inspector at Kennedy Space Center.

When that work ended, I had many quality jobs around the KSC area. I currently support the build of the Orion Multi-Purpose Crew Vehicle as a NASA contractor quality assurance specialist.

My aerospace career has had many winding paths and all of them were possible because I was willing to take some risks and get out of my comfort zone. The one thing I would like to emphasize is to never give up on a dream!

Brooke Edwards

"A bright future in space exploration was dawning, and I sincerely want to be a part of it as a space science communicator."

— BROOKE EDWARDS

Growing up in Philadelphia, my life was not easy. Living in a small middle-class neighborhood, many of the other children were fortunate enough to have the experience of travel. Meanwhile, life outside the city was barely experienced. Not even the Jersey Shore, a destination only about eighty miles away, where literally everyone flocked to in the summertime. My grandmother did her absolute best to provide for me, yet her fixed income and declining health kept us limited to the surrounding neighborhoods. She encouraged me at an early age to work towards going to college and even dream of being an engineer!

I was a bright and curious child, but my talent laid in the arts. If not drawing cats or making clay animals, elaborate fantasy stories were being written. Even at the age of three years old, I could sit and write a fantastic story for over an hour.

Around the age of three years old is when my first love came into play. Even twenty-eight years later, the memory of the night feels like yesterday. There is no recollection of where we were traveling that night, yet it all started when we headed down the alleyway behind the rowhomes. As I sat in the backseat of my grandmother's Plymouth, it appeared we had a stalker. Above the houses to the left of me was a bright crescent moon. Everywhere we went, the moon seemed to follow us. Curious, I questioned why we didn't lose sight of the moon. The answer I received was fantastic, opening my eyes to the fact that a whole universe lies beyond Earth. My imagination was grabbed and never given back. I fell in love.

Later on, the evening news and books proved there were people known as astronauts, who were fortunate enough to explore the space outside of Earth. Wonderful news footage showed astronauts flying in shuttles. Laying in a huge cardboard box one day, I pretended to be flying on the shuttle. One would say that is a common fantasy for children. Just what happens if that fantasy becomes a dream that doesn't quit? Every time I saw my first love in the sky, it was a reminder of the dream that was pulling at my heart.

It is when you get to middle school that reality begins to knock you down. Being a child more talented at the arts, my poor math abilities made it clear I would not be an engineer. This fact combined with other factors made it clear to me that I would never be a NASA astronaut, which was a secret ambition I kept to myself. My curiosity for natural sciences alone was not going to launch my desire, yet I kept on reading the space chapters in my science textbook long before they were even covered in class.

Going through high school with a shattered dream, I continued to read space news articles online. The topic of space tourism was discussed, yet in my heart I did not want to be a "space tourist". Desiring much more than simply flying to space, I wanted the job of an astronaut, which I believed was to share the experience with others. Also on my mind was the fact that I had hardly left the tristate area. What were the chances I'd ever get that chance or even afford it when I was older.

In college, my path was settled on the next best thing; studying education and natural science. Putting many hours of sweat and tears into my brutal-yet--fascinating science courses, I refused to let my "mathematical handicappedness" stop me. Even as my grandmother's health began to decline, and finding myself caught in a long-distance relationship, the battle raged on. The day Hurricane Sandy slammed into the East Coast and straight into Philly, I was glued to my chemistry book, determined to pass. The eerie daytime darkness and sound of the windows rattling is still strong in my flashbacks. That storm cleared, and a year later saw me passing and graduating. Yet, happiness was short-lived after graduating into the "real world".

The real world saw me struggling to find a job while caring for my grandmother, whose health was failing. After finding stable employment, it was finally possible to visit my boyfriend of five years in Michigan. This was the first time I had ever flown, and the experience of taking off into the sky instantly amazed me. Hooked on this experience so many take for granted, I wanted more. My dream of spaceflight was fueled.

A few years later, something interesting struck excitement into my grounded dream. An organization called Space for Humanity was seeking astronauts to fly on future commercial flights. Commercial spaceflight was about to take off, and the fact that I might actually have a chance to fly sparked me into action. Even being very introverted, there was an impulse to get in front of a camera and profess my desire to be an astronaut and communicate why space exploration matters.

A few months later, I found myself lying in bed mourning the loss of my grandmother. The strong woman who had encouraged me to achieve greatness was gone. Being a caregiver for years, I was lost.

Lying in bed after crying for hours, it finally hit me; there was something inside me pushing me to pursue space communication. With no idea where to even start, or how to make this a career, I followed my heart. It was leading the way and telling my head to step aside.

After assisting a startup space-media company, I purchased my first astronomy binoculars. Learning astronomy in Philadelphia was difficult as light pollution drowned out much of the stars. A local astronomy club assisted me in my efforts, and eventually led me to volunteering to do astronomy outreach during the 2018 Philadelphia Science Festival. The amazement on the faces of the adults and children I spoke with that night assured my new trail was one to stay on.

It was with a leap of faith that I traveled solo to Huntsville, Alabama to achieve another dream. Adult space camp was all I imagined it to be and more. The experience itself, plus the amount learned about human space was immeasurable. The best part; I lived out my childhood dream of being a mission specialist onboard the space shuttle!

A few more space experiences followed, and I began to keep a personal blog to share about all of my experiences and some space science that would interest everyday people. The main theme was to share space science as well as my personal journey to break into the space world. My social media presence followed shortly after. In love with this space project, I wished there was a way to monetize it.

Eventually I moved to Michigan in late 2018 to be with my boyfriend and within the first few nights, I saw the Milky Way for the first time. A truly dark sky followed months later and left me breathless. The pure sky was a glittering sea of stars.

The next few months found me volunteering to host local stargazing parties at the beach, where once again people were brought to the wonder of the night sky. I also found myself writing space articles for a local paper, once again blending my two passions: writing and space.

Through a few online sources, it became evident that science communication was actually a career. What I had been searching for all these years was right in front of me, and I was absolutely unaware. After rebranding and launching my blog once again, the perfect name came to mind. Thinking back to that pure night sky, it was not "A Sea of Stars," but "A Brook(e) of Stars." There are many more stars out there that are not visible from Earth. One day, when above the atmosphere, I will be able to see them better.

Not only the stars, but also able to look back on Earth and all I take for granted. This was realized in full force when fortunate enough to be selected as an analog astronaut. In February to March of 2021, I served as the science communication officer on a simulated lunar mission that took place at HI-SEAS (Hawaii Space Exploration and Analog Science), a simulated space habitat on Mauna Loa. Our crew was Selene III. Not many people in their wildest dreams could ever imagine making it to Hawaii, let alone explore a tube created by once-molten lava to learn about features that also lie on the moon and Mars!

The mission was challenging. As if truly in space, most communication from those back on Earth is gone. Temperatures were low, resources such as food, water, and limited supplies were low, power was low in bad weather, and one could not leave the habitat for a spacewalk without permission from mission control. This was close to visiting my first love, yet who knew a relationship could be so difficult. The moon was cold, unforgiving of error, came with strict deadlines, and absolute quiet. The only voices I had the luxury of hearing were those of my five fellow crew members. The contact with mission control was through a simple email client with a thirty-second delay. While many would feel tortured, I somehow thrived.

As the door to our habitat opened for my first EVA (moon walk) after being locked inside for ten days, I breathed excitedly inside my space suit. The volcanic rock terrain felt foreign under my feet as I proceeded forward and upward in elevation. Encountering absolutely zero plant or animal life on this expedition, I truly felt as if I was on another world. Every minute was awe-inspiring, and sparked new curiosity into me. It was once I returned to Earth that the impact that the mission had on me was evident. My increased ambition and bravery was noticed by many. Now if someone asked me if I would travel to the moon, I wouldn't even hesitate to say "Yes!", an answer I would have thought twice about just a year prior.

With all the excitement, is it truly the friends you make along the way? During the 2020 pandemic I had taken Loretta Whiteside's SpaceKind class online, which uses her book, *The New Right Stuff*, as a guide to bring your best self forward. After returning from HI-SEAS, I found a small group of us were still in contact. To this day, we still have our weekly call.

The past year has found me exploring Lake Michigan as a newly certified open-water diver. For a whole week after work, I submerged myself in a weightless, underwater environment. As I learned new skills, new discoveries were made and curious lake creatures were encountered. My desire for new adventures and experiences was high. Ready for a challenge, I made the trip that changed me again forever.

Never seeing a rocket launch from closer than 200 miles away, or let alone a crewed launch, I knew this had to change. Inspiration4, the first all civilian mission to space, was my chance. Everything seemed to align seamlessly to get me there and throughout the trip. Traveling solo once again, and to yet another state I had never been to, my mind was strangely calm.

Two nights later, the launch went off perfectly on time. Standing on the beach in Cocoa Beach, FL, I faintly heard the rumble of the boosters as the Falcon 9 rocket flew over me. Speechless over the brightness of the sky and the glow of the rising rocket, I just laid on the sand and watched.

As the Crew Dragon flew into the twilight, forming a beautiful aurora in the sky, I felt in my heart I was in the right place at the right time. The boosters glided down to Earth over the Atlantic Ocean, sending off a distant roar. The little dots of light landed on pads out at sea, and four non-traditional astronauts were orbiting the planet. One day, it could be me. A bright future in space exploration was dawning, and I sincerely want to be a part of it as a space science communicator.

Rachel Zimmerman Brachman

"I am lucky to be part of these journeys of exploration, and I invite you to join me in exploring the planets in our solar system and beyond."

— RACHEL ZIMMERMAN BRACHMAN

International Space University Master of Space Studies 1997 – 98

Strasbourg, France

grew up in London, Ontario, Canada. I always knew I wanted to study science and I was excited to seek out answers to questions about how the universe works.

My parents and my teachers sparked my interest in space. I wanted to know everything about everything. In grade two, my teacher taught me about the planets in our solar system and took my class on a tour of the observatory at the university in my city. I saw Saturn through the big telescope for the first time when I was seven years old.

My parents took my little brother and me out to the countryside to watch the Perseid meteor shower every August. We would see so many meteors streaking across the night sky. I loved the vastness of space.

I participated in science fairs for twelve years when I was in elementary school, middle school, and high school. My science fair projects were displayed at my school science fairs in elementary school and middle school. My high school didn't have a science fair, but I did science fair projects anyway. From grade six to grade thirteen, my science fair projects competed at the London & District Science & Technology Fair. My projects in grades seven, eleven, and thirteen went on to compete at the Canada-Wide Science Fair in Cornwall, Ontario (1985), St. John's, Newfoundland (1989), and Vancouver, British Columbia (1991) so my projects gave me opportunities to see parts of Canada that I wouldn't have seen otherwise. I'm still good friends with some of the people I met through the science fair.

My grade seven science fair project was an invention I created and programmed, which enabled non-speaking people to communicate through writing. I tested my invention, the Blissymbol Printer, with children and adults who had cerebral palsy. The project went from my school science fair, to my city's science fair, to the Canada-Wide Science Fair and then to the World Exhibition of Achievements of Young Inventors, which was held in Plovdiv, Bulgaria. Through this project, I joined the Women Inventors Project and the Women Inventors Networking Society.

In high school, I had summer internships in research laboratories, studying rehabilitation engineering, mechanical engineering, cancer research, and astronomy.

I studied physics at Brandeis University in Massachusetts, and I took history of science classes, including a slide lecture course called Images of the Cosmos and a course on Women in Science. I also took a neuroscience course called Man in Space, which covered a brief history of space exploration as well as the physiology of what happens to the human body when it's in space. I participated in an experiment in a slow rotation room to see if eye level appeared to be different when spinning with a force of 1.5 times the force of Earth's gravity.

I earned a Master of Space Studies degree at the International Space University in Strasbourg, France in 1997 – 98. I was the class president, which gave me an opportunity to work on my leadership skills. I had an internship at NASA's Ames Research Center in Northern California where I studied medical and rehabilitation applications of space technology. I worked with an organization called the Tetra Society to draft an agreement that enabled NASA employees to volunteer a small portion of their time and expertise to build custom assistive technology for disabled people in their community, so that people with disabilities could have an improved quality of life.

After I graduated from ISU and moved back to Canada, I went back and visited my grade-two teacher who had taught me about the planets many years before. She was in her last year of teaching before she retired and she invited me to talk to her young students about what I had learned in space school. I ended up visiting her classroom three days a week during the month when she was teaching her space unit. I talked with her students about rockets, astronauts, and exploring the moon and other planets. We even went outdoors on a snowy day and built a snowball solar system on the school's playground.

I worked as a contractor at the Canadian Space Agency in 1999-2000, where I built the black-and-white target dots that are on the outside of the International Space Station (ISS). The dots were used by the video camera on the end of the Canadarm, or Remote Manipulator System, that was used for assembling the components of the ISS as they were launched to orbit on space shuttles and russian rockets.

My next job brought me to Pasadena, California, where I worked on education and public outreach at The Planetary Society, an organization that was founded by astronomer Carl Sagan, former Jet Propulsion Laboratory (JPL) director Bruce Murray, and former JPL engineer Louis Friedman. I ran international essay contests and art contests for students around the world.

Next, I worked at the California Institute of Technology (Caltech) at the Center for Neuromorphic Systems Engineering, a group of ten professors and their graduate students who had funding from the National Science Foundation to design and build biologically inspired robotics. They made robots that could swim like a squid, hop like a frog, slither like a snake, and flap their wings like a bat.

In 2003, I started working at NASA's Jet Propulsion Laboratory in Pasadena, California. I worked on education and public outreach for the Jupiter Icy Moons Orbiter, which never ended up being built. I worked in JPL's Education Office, running summer internship programs for minority students and high school students. In 2006, I started working on the Cassini-Huygens mission to Saturn and Titan, which I worked on for eleven years until the end of the mission in 2017. I ran international essay contests about Saturn and its rings and moons.

In 2018, I worked on citizen science projects at JPL about the Earth (the GLOBE Observer program) and Exoplanets (Project PANOPTES). PANOPTES let people build their own telescopes to study planets beyond our solar system.

Now I work on radioisotope power systems public engagement, discussing the power technology NASA uses to help its spacecraft explore places that are too dark, too distant, or too dusty to use solar panels. I work as the technology transfer liaison between JPL and NASA Headquarters. I also work on a project called Exoplanet Watch, which is a citizen science project that lets anyone and everyone study planets beyond our solar system, by observing changes in brightness of a star as its planet passes in front of it. You can use your own telescope, if you have one, or use data from other telescopes, to learn more about planets that are hundreds or even thousands of light years away.

I love being able to share my passion for space exploration with other people. I applied to be an astronaut several times, but now that I know more about robotic exploration of space, I'm even more excited about what we can study by sending robots to be our eyes and ears in space, exploring planets that are much farther than we'll be able to send people anytime soon.

The twin Voyager spacecraft have been in space for forty-five years and are now in interstellar space, the space between the stars. Rovers like Curiosity have been on Mars for a decade, teaching us more about Mars than we've ever known before. My colleagues at JPL are building the next spacecraft that will travel to Europa and map the whole icy moon, giving us views of this habitable ocean world that we've never seen before.

It is an exciting time to be studying the planets in our solar system and the exoplanets that orbit stars beyond our own Sun. I am lucky to be part of these journeys of exploration, and I invite you to join me in exploring the planets in our solar system and beyond.

Sarah Treadwell

"...it is such an exciting time to get into STEM and there are so many ways you too can get involved!"

– SARAH TREADWELL

When I was a little girl, I asked a lot of questions. Sometimes this got me into trouble. But I didn't let this stop me. I also had an adventurous spirit and moved to lots of different parts of the country. I was very brave and lived my life the way that I thought I should, even though that went against what I was taught growing up. What really changed my life, however, was becoming a mom and having a daughter of my own. I realized that the only way I could encourage her to live her life to the fullest was to show her how to do it with my own life.

Soon after Adaline was born, I bought my first telescope. The first time I looked through it at the moon, I knew I needed to share it with other people. I didn't know yet at the time, but I was doing something called "science communication" through sidewalk astronomy and this was the beginning of my space science career. I joined telescope clubs and started learning all I could about astronomy and space.

Then I learned about something called "astrobiology". Astrobiology studies where life lives on Earth in really remote and extreme places to better understand how to look for life in our solar system. I thought it was the coolest science I had ever heard of! I learned about people like Carl Sagan and the Voyager missions with golden records that sent messages from us on Earth to space. I thought it was the coolest science I had ever heard of! But I had a bachelor's degree in English and didn't think it was something I could pursue as a career.

One night, I was coming home from an astronomy event and I got into a very scary car accident. The accident was like a light switch that went off inside me. I realized my life was too short and precious to not follow my passion, no matter how impossible it seemed. The next day, I started calling around and asking people how I could get into astrobiology, even though I hadn't studied science in college. I started taking science classes at a nearby community college and planned on applying for a master's degree in Biology.

Then Covid-19 shut the world down. I was very worried that this was another hurdle that may slow me down, so I started looking for opportunities to keep working towards my dream. I found an online internship at the Blue Marble Space Institute of Science in science communication and astrobiology. It hadn't occurred to me until then that I could pursue a career in learning and sharing about all the science that I found so cool! I could have a career in talking!

That internship was the door that has opened so many more doors. From there, I was asked to join the production team of "Ask an Astrobiologist", a show that airs on NASA TV. I went to the Mars Desert Research Base as the station journalist and got to experience an analog astronaut mission. I learned how to SCUBA dive to simulate the astronaut experience and that has led me to many more adventures, including going to Mount Everest base camp at an elevation of 17,500 feet! That's as close as I've gotten to space so far!

I realized that science communication was where I was meant to be and started a master's degree in communication. More and more opportunities kept opening up for me, including going on a two-month astrobiology expedition on board the JOIDES Resolution as an on-board communications officer. At the time of writing this, I leave for the ship in one week. This ship is a floating lab and we are studying the Lost City Hydrothermal field, the most interesting place on Earth for astrobiology!

If there is any advice I can give to young girls looking to get into STEM, it is to just say *yes!* And don't be afraid to explore creative ways to get involved. Saying yes has given me the chance to write articles, host a radio show, produce live stream shows with top level people at both NASA and the private space companies. I just now got into a PhD program for communications! I love what I do and put my heart and soul into everything I try and apply to and that passion is evident to everyone I engage with.

Despite all that I do, my favorite thing to do still is take my telescope outside and show people the moon. Sharing the moon was how I got into science communication and I truly believe it is the best way to get other people excited about space exploration. Now, as we look to send people back to the moon for the first in many of our lifetimes, it is such an exciting time to get into STEM and there are so many ways you too can get involved!

Janelisse Morales

THIS is what a rocket scientist looks like.

— JANELISSE MORALES

Hello, I'm Janelisse Morales Gonzalez and I am currently a systems integration/test engineer working on the Orion program. Ever since I was six years old, my parents took me to the Johnson Space Center in Houston for the first time to look at the Saturn V Rocket. I remember looking at the engines directly and thinking to myself, *I want to build something like that one day.* Never in my life would I think I would have the chance to work for NASA, Blue Origin, and now Lockheed Martin.

It took so much hard work to become an engineer. First it was a calculus professor telling me I would never be able to work for NASA one day (before I got a NASA internship), then came the constant studying and bad test grades, and sadly my self-esteem lowered and I thought I could never become an engineer. But at the same time, I couldn't let my six-year-old self let go of her dreams. So I pushed on and continued to study hard until I graduated college at twenty-one.

As a young, soon-to-be twenty-three-year-old Puerto Rican woman, it is hard to be in a male-dominated industry, but I love what we are doing with this book and I know it will impact many girls' lives.

Following, I have two LinkedIn posts that went viral:

Fourteen years ago, I was a six year old who had a goal of working for NASA one day.

Six years ago, I began my aerospace engineering degree track at my community college.

Three years ago, I was bullied by my professor for wanting to work in the space industry and was told it was "impossible" because of both my ethnicity and gender.

Two years ago, I was accepted to University of Central Florida, however I began questioning my future and asked myself if engineering was really worth it.

And about one year ago, I applied to over 150+ internships with nothing in response but rejections, except for one internship offer I received two months ago.

Today, I start my first day as a NASA intern.

Twenty-one years old. Hispanic. Female.

These are the characteristics you probably wouldn't think of if you try to picture an aerospace engineer, but here I am.

I've told this story many times, this dream all started when I first saw the Saturn V rocket at NASA's Johnson Space Center at six years old and I haven't looked back ever since. With a lot of perseverance and determination, I am so proud to announce that I will be graduating with a Bachelor's in Aerospace Engineering in May.

THIS is what a rocket scientist looks like.

Carla Tamai

"...trust me, the view from the top is, in the end, very worth it."

– CARLA TAMAI

Hi. My name is Carla Tamai. I am twenty-four years old. I come from the south of Italy, Naples. Ever since I was a child, I have always been very curious about anything that I studied and learned. In every assignment I got, be it important or not, I felt like I wanted to do a good job and I used to dedicate myself entirely to it. In high school, I studied at a "Classical" lyceum, where mainly humanitarian subjects were taught – such as Latin and Greek literature and grammar, Italian literature, history, and philosophy. There were also subjects such as mathematics, physics, chemistry, art, Catholicism, and physical activity, but they were less frequently taught. Studying those subjects allowed me to think out of the box, taking into account all possible points of view.

It also helped me very much from an organizational perspective. It was a very tough and intense experience, but it taught me how to set boundaries between work and leisure activities. It has helped me approach different problems, decisions, or even random conversations with strangers. In my last year of high school, however, I had to decide on my upcoming university path, and I was in real trouble. I liked everything. I wanted to study mathematics, physics, languages and linguistics, law, and medicine, but obviously I couldn't do it all.

So, I took some time for myself during that period to narrow down the options to at least two to three of them and I came to the conclusion that I was mainly interested in the fields of mathematics and physics. With this in mind, I chose a career in aerospace engineering to combine both areas of expertise. Even though I desperately wanted to study abroad and experience life outside of my home country, I stayed in Naples and did my bachelor's at the University of Federico II. It is among the best universities in Italy, with very competitive courses for scientific subjects such as engineering and therefore many people were questioning my decision.

Since I was coming from a humanitarian background they believed I did not have enough scientific knowledge and expertise to proceed on that path. However, looking back, for me, that was the perfect one. At the end of my bachelor's degree, I got knowledge from both a scientific and a humanitarian point of view, and I won two scholarships for outstanding grades. The mindset I got from my high school helped me to face very diverse subjects of the bachelor. I enjoyed every single class on aerodynamics, structures and materials, systems engineering, pure physics and mathematics, computer science, and more. However, as you can already imagine at this point in the story, during the last year, I felt like I still did not find my true passion. Be it for gender prejudice or personal reasons, engineering seemed to not perfectly fit with me. One thing was sure though, I wanted to pursue my career within the space sector. The more I got to know about the space field, the more fascinated I was.

There were many aspects I enjoyed studying such as planetary sciences, astrophysics and astrodynamics, exploration, orbit determination, and mission design. I started thinking about it as my passion, but at that time, I had no idea how vast this field is, and what it really meant to work in those fields. This led me to finally choose the Space Exploration (Space Flight) Master at Delft University of Technology. It is not only among the best ones in Europe, but it also allowed me to do the experience abroad that I was very much willing to do from an early age. Therefore, as soon as I graduated from the University of Federico II in Naples in Aerospace Engineering, I flew to the Netherlands with as much enthusiasm as a little kid during Christmas.

However, it was only after a while that I realized that educational systems in Italy and the Netherlands are very different and that it would have been harder than expected to get used to that new system and environment. Even though I waited for that moment for a long time in my life, those two years in Delft were very tough. During that period, I learned about many different topics but, since I had to manage multiple deadlines, lectures, and exams at the same time, I hardly had the chance to think about what would have been interesting to me for future job careers. I lost a bit of interest in everything, and I refused any kind of job that was being offered to me as soon as I graduated.

In the end, I found myself with the best graduation grades from both Universities and yet a bit of discomfort. I felt I did not like most of the jobs I was being offered and that they did not represent the future job that I wished to have, even though I still did not know what that would have been about. It was at this point in my academic career that I could have easily given up. But I did not. I did not want to settle for a job that I did not like, so I kept looking for other study opportunities, to hopefully offer me what I wanted. In August 2023 I started a second Master's degree at the International Space University (ISU). There, every single aspect of space is taught – from space policy and economics and law to management and business, human performance in space, applications, humanities, science, and engineering. I believed there was an enormous part of the space sector that I was missing out on. I wanted to expand my knowledge and find what was my passion.

Turning down those jobs before ISU was not the easiest thing to do, especially after two years of living abroad and not being self-sufficient. However, that was also the best decision I have made so far in my life.

It was a turning point in my career and life. At ISU I found what really thrills me, the Human Space Flight Operations, and I learned how to fight for it. I kept studying that year, becoming more motivated and believing in myself and my capabilities. Whenever someone says "Smile at life and it will smile back at you back…"

Well, that is very true. Knowing my strengths and my ultimate goal, I received more job offers that really interested me, among which was the European Space Agency (ESA) – the dream of every space enthusiast. I am now completing my internship at the European Astronaut Center (EAC), which the Space Medicine Team and I will soon start in September 2023 as a Young Graduate Trainee (YGT) in the same ESA center. The project I am going to work on as a YGT is gateway training analysis, planning, and facilities definition. I am very excited and looking forward to starting.

My takeaway from all of this is that daring to change and chasing your dreams does pay off! There is no right or wrong in what we decide for our own life, so if I might give a suggestion to anyone reading the book and these amazing stories is to dare. Dare to ask, meet people and explore. Do not be shy, especially if you are a girl. Be brave enough to leave your comfort zone, even if no one believes in you. In the future, you might want to change your career path, be brave enough to do it, and then prepare yourself to *enjoy* the ride.

Not everybody is born with a talent, and it is normal to not know what to do in life at a very early stage. Give yourself time to understand what to do, and do not stop trying different things if you haven't found what thrills you. Be curious, because that will allow you to discover what you like. Sometimes people are born with their future already written, but in most cases, we have the opportunity to shape and determine our paths. Never take it for granted! It will not be easy, the road can often seem very uphill, but trust me, the view from the top is, in the end, very worth it.

Good luck to any beautiful, strong girls out there in the world! And please reach out to me if you would like to have a chat!

Sharon Caples McDougle

" There are many career paths in the space industry – you don't have to only aspire to become an astronaut. "

– SHARON CAPLES McDOUGLE

My name is Sharon Caples McDougle and I am writing to you to tell my story and let children know that they can achieve whatever they want in life!

I grew up in a small town, Moss Point, Mississippi with eleven brothers and sisters. My dad died when I was four years old and my mom died when I was seven years old and I was very sad without them. I lived with my oldest sister and her family until I graduated from high school. I had to work very hard at home because I had to do all the chores by myself.

I loved and looked forward to going to school every day – I even wished I could go to school on Saturdays and Sundays! I loved learning new things, and I really loved reading and spelling. I won the spelling bee when I was in second grade and was chosen to read "Twas the Night Before Christmas" for our Christmas program. I remember when I was too young to go to school, I would look through my brothers' schoolbooks, dreaming of the day I'd be able to go to school too. I had the best teachers, and they played a large part in shaping me into the person I am today. They not only taught me reading, writing, and math, they also gave me lots of hugs and praise for getting good grades and I really liked that.

I wanted to go to college and become a kindergarten teacher or be a flight attendant when I grew up, but I didn't have money to go to college. An Air Force recruiter visited my school and spoke to my class in my senior year of high school. He said we could travel and see the world, and they would pay for college while working in the Air Force. After graduating from high school, I joined the Air Force and served for seven and a half years.

I attended the School of Aerospace Medicine where I received my training to become an aerospace physiology specialist and was the only African American woman in my class. I assisted in high-altitude training and performed hazardous duty as an inside observer during hypobaric and hyperbaric chamber operations. I also inspected and maintained flight equipment used by pilots/aircrew in the SR-71 and U-2 reconnaissance aircraft ("spy planes"). The equipment included full pressure suits, survival equipment, and oxygen systems. I sized and fitted the pilots/aircrew members' pressure suits, suited them up, and tested them before taking them out to the aircraft and strapping them in. I also loaded the survival seat kits and parachutes onto the aircraft and strapped the crewmembers into the aircraft. I traveled to several countries, including Greece, Korea, Japan, and England, as well as stateside locations, many times in support of the aircrew and missions. I had the best job in the Air Force. I had so much fun traveling the world that I never went to college.

After working with pressure suits in the Air Force, I realized the only place I could probably get a job would be at NASA working with the astronauts. I worked as a Space Shuttle Crew Escape Equipment (CEE) Suit Technician. I was the only African American team member in my department at the time I was hired. My Air Force training and experience played a major role in me being hired. All astronauts who flew aboard the space shuttle had to wear the launch/entry suit (which was changed to the Advanced Crew Escape Suit later). The equipment they had to wear included the suit, helmet, boots, gloves, communications cap, harness assembly, liquid cooling garments (underwear), and most wore diapers too.

The highlight of my career was suiting up Dr. Mae Jemison. Everyone knows Dr. Jemison was the first African American woman to travel into space, but many don't know that I, an African American woman with her own firsts, suited her up. I was Mae's suit tech for her historic mission aboard the space shuttle Endeavor. I worked closely with her during all her training leading up to launch, as well as actual launch day and landing of the space shuttle. I inspected and tested all her equipment to make sure everything was in the best condition for her training events and launch day. I would prepare everything for her and help her put on all her equipment.

After working there for four years, my supervisor noticed I was performing crew chief duties while I was still a technician and promoted me to crew chief. I was the first woman and first African American to be promoted to CEE crew chief. I worked very hard; I was reliable and dedicated to my crew and the company – earning the respect of my peers, astronauts, and management. In my new position, I was responsible for leading a team of technicians to suit up astronaut crews. I had to make sure the astronauts were provided with outstanding support during suited training, launch, and landing events. I traveled from Houston to the Kennedy Space Center where I worked in support of many space shuttle launches, suiting up many astronauts. One of my most memorable missions as crew chief was leading the first and only all-female suit tech crew supporting the space shuttle mission STS-78.

Later in my career, I became the first and only female and first African American promoted to the position of Manager of the CEE processing department. In this position I managed a team of twenty-five plus employees responsible for processing the equipment worn by the astronaut crews aboard the space shuttle and strapping the astronauts into the space shuttle. I held this position until the space shuttle program ended in 2011. I continued working until 2012 to help close out the program, ending an illustrious twenty-two-year career with the space shuttle program. I was very fortunate to have one of the coolest careers ever.

In closing, I'd like to leave you all with this: Anything is possible, if you work hard to achieve it. Don't be afraid of technology, science, or space exploration. There are many career paths in these fields – you don't have to become an astronaut. I also advise you all to not put off going to college if that is the path you choose. And, lastly, you can have a successful career without having a college degree.

Eleonore Poli

"You have this gift, this gift of believing in something greater than yourself... Ad astra."

– ELEONORE POLI

Have you ever watched a rocket launch in an auditorium, surrounded by fellow engineering and physics students, with the floor trembling from the sound of the blastoff and tears in everyone's eyes? Have you wanted to sit on that rocket, on that engine with the energy of a bomb, and to be propelled into space ?

In March 2022, I got the chance to see Didier Queloz, one of the Nobel Prize winners for the discovery of exoplanets, at Fantasy Basel, the Swiss Comicon. He gave a talk on exoplanets while covered in paint for Comicon and smiling from head to toe.

This is the space community. It brings any kind of professional, from artists to physicists and engineers to managers together to talk about space and to share about new developments, new images such as the ones provided by the James Webb Telescope. This community is wonderful, and I will tell you my story – how I got to hug an astronaut, how I moved to Soyuz, and applied to become an astronaut.

My name is Eleonore Poli. Although my mom was French and my dad was Italian, I grew up in Switzerland and I'm a Swiss citizen. Space isn't as popular in Switzerland as it is in some other countries such as the USA, however I was interested from a very young age. I loved engineering and stars. I was fascinated by machines, especially submarines. At the age of seven, I wanted to solve climate change and become an environmental engineer. I tried to draw a few concepts, such as a robot reminder for the elderly to drink enough during heat waves and a turbine working with water currents produced by underwater volcanoes. They were all squibbles, but I loved the creativity of engineering. Stars were a dream to me. I was also interested in becoming an astrophysicist. The job of astronaut, although interesting, seemed totally out of reach – the astronauts seemed too perfect. They were smarter, stronger and just better than me so I put it out of my lifeplan as a child.

At around thirteen, I became more and more fascinated by planes. I attended airshows and learned aircraft properties by heart. In high school, I wrote a book on how to build and fly aircrafts and on aviation history. This was where I started my first aerospace family: I met the organization hepta.aero, which specialized in student-based projects on airships, reverse engineering of aircrafts, and discovering abandoned aircrafts.

By the age of seventeen, I had flown a Boeing 737 simulator, had a 50cm plane model in my room, and registered to become a Swiss military pilot. I still wanted to become an engineer, but I enjoyed the idea of playing with a 3-million-dollar toy in the sky while not actually shooting at anyone. I did not get in, but it did confirm my thirst for aerospace and flying.

After attending some university faculty presentations, I realized that being an astrophysicist would be a tough job with a small chance of reward and not necessarily something I would excel at. Environmental engineering was a disappointment. It did not focus on building machines and focused more on monitoring and simulations. Luckily on these presentation days, I attended twenty presentations, from graphic designer to mathematicians, civilian pilot to physicist. I also attended a material science and engineering presentation. And it clicked. Although I am not diagnosed with ADHD, I have so many areas of interests that you might think I do. I have played the piano since I was three, I have done over ten different sports, and I train ten to fifteen hours a week. I love photography. I even worked as a robotics teacher assistant as a teenager. Material science was the answer to this spread-out attitude. With this field of study, I could work on planes, satellites, sports shoes, and piano parts. I could do anything.

I enrolled and started the curriculum. I wasn't the best student, but made the best of the campus by attending any conferences on astrophysics, space, and planes that I could. I became a member of the student aerospace association, and my aerospace family grew. Every project and thesis I had, I would orient it on the theme of aerospace. I researched polyurethane enveloppes for airships, then 3D printed aluminum silicium open cell foams for satellites for my bachelor thesis. I worked in material science laboratories and as a math and robotics assistant to make money.

But between an overly filled schedule, a difficult course, and then an unfortunate family event during my exams, I failed my curriculum. At the time, I was just starting my internship at Pilatus Aircraft Ltd. This Swiss aircraft company is located between two mountains. They produce and test the aircrafts there between cow excrement and wonderful lakes. I felt ashamed. Ashamed that I had failed the people employing me there and that I was not as good as they thought. I thought I was a failure and, although I loved my job there – qualifying brazing alloys for the company – I was dreading their judgment.

On a hot day during the summer of that internship, I had absentmindedly removed my shoes under my desk when my boss walked in and asked to speak to me. Panicked, I thought he was scolding me for my shoes, but he closed the door of his office and laughed. He then looked at me kindly and said he had learned I had failed my degree and asked if I was ok. He told me there were other solutions, asked me if I will take a loan for studies, and shared his own experiences. It was such an unexpected, wonderful, kind, and human thing. It comforted me that I could still do a great job – that people trusted my work, not the diploma.

I proceeded to continue my studies in Swiss Germany, doing mechanical, materials, and process engineering in German. I would go to bed at 5pm for two weeks as German was my third language and it was completely draining. But I completed my degree, did some Russian lessons, learned a new sport, and did some extra classes on space mission design. I even met the director of the Swiss Space Museum and bugged him enough that he proposed I volunteer for the next exhibition. My space family grew exponentially: I met space nerds, taught people to use an ISS VR simulator, and explained space facts.

Feeling better about my failures, I applied to aerospace masters and material science programs in the UK. I applied to only three universities. The third one was a joke application. I did not think I would get in. Yet, a month later, I received a series of emails. The University of Cambridge had accepted my application and was offering me a PhD position in materials and metallurgy. As the email was on a Sunday evening at 10pm, and I hadn't even graduated from my bachelor yet, even less from a masters, I did think it was a joke.

However two interviews and one graduation later, I was starting my masters and PhD program at the University of Cambridge. I first worked on the thermal barrier coatings (heat shields) for turbine blade applications. I was operating a submarine-like machine (imagine my delight!) which created a vacuum, where a plasma was struck and ceramic powder melted to form a coating. For my PhD, I then worked on hot corrosion resistant coatings, also for turbine applications. Being in vibrant Cambridge, I met incredibly motivated people, people you could ring up at 4am to go for a run, build a start-up with, or listen to a concert and wrap up the day at the pub. I read more and more books from astronauts on astronauts. The light was back. I was going to aim to become an astronaut again.

It got really real when I read astronaut Scott Kelly's diary-like endurance book on his year-long journey in space and the events that led to it. Spoiler alert: Scott Kelly is a successful astronaut. This you know before picking up the book. It is the reason why you pick it up. However, what I did not know was that Scott Kelly wasn't a model student. He got his dreams crushed. He had to work hard. He didn't have the perfect upbringing. I read his book on a plane, and felt full of love, sadness and hope. If he had failed and made it there, then I should try. Regardless of my failures, I should try no matter what.

The journey is what makes life interesting. The fight for what you believe in is what gives you shape. The achievement is the cherry on top. But the real gift is what you've learned along the path.

Into my second month of PhD, which was filled with beautiful and unfortunate events alike, I saw an advertisement by the Swiss Space Centre: "Astronauts wanted." Right away, I threw away anything I was doing, and dug into the ad. They were looking for students to simulate a lunar mission, an analogue space mission. I was the first candidate to send a CV and a motivational letter. The recruitment was based on a similar fashion to that of the European Space Agency, a CV plus a cover letter, then a video for English skills, plus a motivation assessment (intrinsic/extrinsic), then cognitive test, a medical test, and finally, groups tests, a psychiatric evaluation and an interview with astronaut Claude Nicollier.

Professor Claude Nicollier is the one and only Swiss astronaut. As a physicist and test pilot, he worked on the Hubble telescope in the space shuttle missions. I attended any conference he gave and was delighted to have the honor to talk to him in such a context. I was selected with seven others to become the first analogue space missions made for students by students. We became a crew very quickly, starting with extreme environment training such as sleeping in snowstorms, ice diving, and night ice diving with French explorer Alban Michon.

Covid postponed our mission from April 2020 to July 2021. It gave us more time to train and to grow. In July 2021, our mission took place 450 meters under the surface of the Earth 1.5 kilometers into a tunnel under the Grimselpass mountain. For eight days, we followed a flight plan such as that observed by astronauts on the ISS. Our food was monitored – we had an inventory. We performed scientific experiments, exercises and reported on our health. As commander of the crew, it was my job to balance the wellbeing of the astronauts and the progress of the mission. As tiring as it could be, I loved every second of it. This mission was the confirmation that the job I love and the job I want, is that of an astronaut.

Shortly after starting the training to become an analogue astronaut, Covid put everyone into a forced isolation, or a low-key analogue space mission as I prefer to call it. It was during the first month of lockdown that I attended as many online space webinars I could to compensate for my lack of data for my PhD. And I learned about other analogue space missions and realized there were no databases about them – no fluid communication. After some prodding, I decided to organize the first conference reuniting all those analogue space missions entities.

Indeed, although I believe we need to take care of Earth and allocate as many resources to it as we can, space exploration is one of the biggest pull for technology and scientific advancements. However, on Earth we have wars, we have greed, we have a bad allocation of resources. I wish for humanity to go to space as one. To go with a chance to become a good civilization. To learn and progress.

People that participate in analogue space missions are not your average Joes in the sense that they agree to discomfort and a time investment with great pleasure for the purpose of advancing research. They might be your future Martian. It became my duty to reunite all of us on Earth, to become a community, before heading to space all together. If we bond on Earth, exchange, work together, we have a shot at taking care of Mars and the Moon.

I founded CHASM, the conference on human analogue space missions and nagged analogue astronauts, engineers, public resources and anyone I could find. CHASM successfully took place in Cambridge in 2022, co-organised with Mars Society UK, and the large support of the British Interplanetary Society. After leading the project, I decided to extend chasm into three objectives: the creation of a database on analogue missions, a series of workshops for continuous inter collaboration, and of course CHASM 2024.

In the process, I presented analogue space missions in several talks and webinars and became a consultant for expeditions and bases as well as an ambassador. My love for space and analogue worlds increases every day, so I am undertaking extreme environment training for expeditions to the Arctic and Atlantic oceans before hopefully going to space.

I applied to the European Space Agency astronaut selection of 2021 but was unsuccessful (with no surprises with all the experience I still need to gather). I coached colleagues into getting in and now am working on preparing the new opportunities for aspiring astronauts in Switzerland. As a career astronaut candidate for AdvancingX, I hope to secure my ticket for space one way or the other after finishing my PhD. Today, I am close to handing in my thesis. I have taught an entire class on analogue space missions, enjoying every minute of research on the subject and planting easter eggs in my class for my students. I am preparing an expedition to revive an analogue base in the Arctic. I officially joined the Swiss Space Museum as an operation team member and am bringing together the Swiss aerospace sector.

As a career astronaut candidate for AdvancingX, I hope to secure my ticket for space one way or the other. After handing in my thesis in 2023, I started the next day a postdoc in additive manufacturing, allowing me to research metallic components for space, medical and automotive applications. I have taught an entire class on analogue space missions, enjoying every minute of research on the subject and planting Easter eggs in my class for my students. In the process of the course, I enlarged my organisation, CHASM, and we expanded to create a research consortium on parastronauts. This allows us to conduct research on how to prepare missions for all, all genders, backgrounds and disabilities.

We recently organized a conference on analogue missions that was blind, deaf and wheelchair accessible, and it felt like the most empowering and heartwarming event. We welcomed mission organisers from around the world, and I was lucky to participate or help organise missions with them, from the Arctic to Iceland and Armenia. Looking into my future, I will shift focus towards becoming a pilot, both for planes and helicopters, to help perform rescue in the mountains and around my country, while working as an engineer. Swimming in the space sector is a wonderful feeling – with opportunities rising left and right, from mission simulation to engineering of rocket parts, and even Swiss Space Museum exhibitions, in which I've been lucky to join the organising committee in 2022. The path to what we desire is just as important as the goal.

Sometimes I wake up wondering how I'll get everything done and then smile and think: You'll do the best and it will be OK. The journey so far has been a beautiful one. Life is exciting. Hard but exciting. And it is most exciting when you have a dream. So let me tell you – if you have a dream, go for it. It doesn't matter if your path isn't straight. It is hard when others disagree or do not understand. But you are luckier than them. You have this gift, this gift of believing in something greater than yourself. Do yourself a favor – get up, work for it, and love the sweat and grit as much as the smile.

Ad astra.

Skye Schwartz | Astro Skye

"Follow your dreams towards whatever path they may lead down or you may create."

– SKYE SCHWARTZ

April 4, 2023

When we think of the space industry, we all have different views on how to get there. Do we want to become astronauts? Do we want to be ground support? Do we want to be the next engineer to build the big rocket? The truth is, there is no right answer, and there are a lot more paths than you may realize.

I was fifteen years old and clueless about anything in space. I was focused on performing arts, and that was my goal. So, on March 28, 2013, when I went to the Kennedy Space Center for the first time, I was shocked. From that day on, my world turned upside down – literally.

I spent the remainder of my high school career following NASA and spaceflight in any way I could, whether that meant watching NASA ISS Live or being awake at 3am on the couch to watch astronauts blast off from Russia. My goal was to go into the US Navy, become a pilot, and then lead to the all-star title of astronaut. A pretty typical OG path if you want to call it that; not that that's a bad thing, but it was the most typical path, especially back in the original Mercury 7 days.

Now, I sit here ten years later since that very first day I stepped foot at NASA's Kennedy Space Center, and I look back and truthfully am cracking a smile and a slight laugh. Oh, how things have changed.

Life has a funny way of keeping you on your toes. I was always the person to have a ten-year plan and make sure everything was going to go the way I saw it no matter what, but, along the way, I encountered quite a few hiccups, path changes, and challenges that have led me to where I am today, not only career-wise but also shaped me into who I am personally.

After I graduated high school, I started at Embry-Riddle Aeronautical University, studying human factors and mechanical engineering as a double degree. I got the opportunity to work on some amazing projects, including NASA NEEMO 20 and ERAU MEERs, and I have graduated from the NASA L'SPACE MCA and NPWEE academies. I *loved* the projects. They were a healthy addiction of passion and experience within the industry, but I did not like engineering. I went from a 4.0 GPA to a 2.8 GPA in one semester simply because I stopped trying. Now I am not saying to go ahead and do this in college – I learned the hard way – but I made an ill-informed decision to jump into a major I did not like just because it was what all of my mentors and most astronauts had a background in.

Now you're probably thinking, what about the military? Well, over a few years and a lot of trying, I discovered I could not take that path. I faced a lot of hurdles during the later years of my undergraduate, and it took me a while to overcome them. There were a few major changes, and I, unfortunately, had some medical issues that ultimately made me take about eight months off of school. I came to find out a few years later I had been medically misdiagnosed, and my journey took a different path. I had a lot of hard times and struggling conversations to try and figure out which way to go and how I wanted to live my life to the fullest. My ten-year plan was crushed in an instant, and it took a lot of coping to figure it out.

I originally created my Astro Skye social media, hoping to spread the word about space and everything it has to offer. I would have never imagined it would have grown into the monster it is today, and I would have never realized how detrimental it was to my recovery to a new path. Through social media, I have been able to speak with so many people, mentor them, and grow myself. It has afforded me opportunities to speak at conferences and share my love of space with each and every person that follows along. I became a NASA Solar System Ambassador in 2016, and it came full circle when I was asked to give presentations at Kennedy Space Center for three years. Was I still on that original path? No. I was forging a new one.

I continued holding onto my NASA Solar System Ambassador position and was fortunate enough to speak at Kennedy Space Center for years. Through the program, I learned truly through each event and through each person I talked to that I could really make a difference. I wanted to make a difference. I was recovering from the mental burden of my world being turned around, and that was when I decided to create Anchors to Infinity, LLC, or ATI, for short. ATI started as a passion project that I thought I was going to conquer by myself. I had this vision of just increasing STEM awareness and helping others; I would have never thought what would come next would absolutely change my life.

I created ATI officially on February 2, 2020. At the end of March, I applied for an intern position at a space education company. I walked in for my first day of work and ended up having a sit down with the CEO. With my background of helping run my family's real estate business growing up and being involved in the operations of projects, I ended up getting a promotion and became the Operations Manager for the business. Talk about an upgrade. I learned a lot and am thankful for that experience, but I also learned what I wanted ATI to become. Just as I felt I was getting grounded and figuring out my life, the world shut down.

COVID. The pandemic shook every corner of the world. That was a hard time. I'd graduated with my undergraduate degree in December of 2020. Jobs were hard to come by, and the world was still. I ended up resigning and moved back down to south Florida from Daytona Beach. The place I called home for five years. It was tough. I kept in touch with one of my old coworkers, Ryan, and he knew I had created ATI. I would talk about my dreams and my ambitions for it, and I realized I could not do it alone.

Although I was 200 miles away and living during a time when the world was trying to get back up on its feet, I knew I needed help if I wanted ATI to grow to its full potential. In January 2021, I started graduate school and brought on Ryan as my business partner at ATI. Little did I know, this was the smartest move professionally and also personally.

Ryan and I hit the ground running and, by the end of the first year, had a copyrighted logo, a website, and a few events set up for the next year. We knew ATI was going to be a slow growth because we were both in school and still are. Ryan is studying Aerospace Engineering, having successfully earned his bachelors in the study, and is now working on his Masters. He was the logical key ATI was missing.

He has continued to help grow ATI to where it is today – a multifaceted organization focused on improving the STEAM industry through education, outreach, and research and development. ATI may be three years old, but it is still a start-up and still something we work towards every single day while also continuing our education.

I didn't take the path I thought I was going to take. When my path got blocked, I forged a new one, and I have had to do that time and time again. What is my path now? I am an entrepreneur. I have created a company to help create and extend STEM resources to kids and adults globally. I may only be helping one client per year and developing at a slow pace, but building a company is a marathon, not a sprint. I speak from my heart and the lessons I have learned. I graduated in December of 2021 with my Master of Space Studies and completed thesis research in Earth anchoring through architectural techniques to mitigate the psychological and psychological effects on long-duration crew members. I am currently in the first year of my Master of Research Administration degree, learning about the administrative aspect of what it takes to do research at companies so I can better improve ATI.

Along the way, I created two more companies, X9 Consulting, LLC and Skye Elizabeth Enterprises, LLC, where I manage my *hop & thought* blog focusing on mental health and my own passions and creations and my Astro Skye social media and website.

Am I still figuring out my path in life? Yes. I don't think you ever figured out your path fully. There are unexpected twists and turns, but that is what makes life interesting. Remember I told you bringing Ryan on would change my life personally too? Two-and-a-half years later, I realized he was the love of my life, and we haven't looked back since. If you were to tell me he was my person three years ago, I would have laughed in your face, but as I said, life is interesting, and I am a firm believer that everything happens for a reason.

Today, I can confidently say I am an entrepreneur with dreams of changing the world in the space industry and in the medical industry towards outlooks on mental health.

Was it easy? Absolutely not. Am I done? Heck no. I have had twenty "no"'s for every yes, and it got to the point I created my own "yes" with ATI. I created my own path, and I didn't care if it was not the norm. I am not here to follow the norm. I am here to break it.

Follow your dreams towards whatever path they may lead down or you may create.

skye@anchorstoinfinity.com

Ruth Nichols

"...the opportunities in aerospace are endless.**"**

– RUTH NICHOLS

Hello, my name's Ruth. I'm currently a college student double-majoring in astrobiology and mathematical sciences at Florida Institute of Technology. I am an Astronaut Scholar, President of the Astrobiological Research and Education Society (ARES), a researcher in multiple labs and groups, and an analog astronaut.

When I was younger, I was not interested in space. I would think to myself, "Why would someone be interested in something so far away and impossible to reach?" I can't relate to adults I meet now who say, "I've always been interested in space," or "I've always wanted to become an astronaut." In contrast, I more or less stumbled upon the aerospace field by accident. I came to discover that aerospace is one of the largest and fastest growing sectors today and that the opportunities in aerospace are endless. It was these types of opportunities that I happened upon, which got me to where I am now.

I used to do fencing, which is a type of sword fighting. Fencing was actually where it all started for me. My fencing instructor was a physicist and, when he learned I was interested in science, he gave me a flash drive filled with science videos, movies, books, and articles. From that flash drive, I learned about something called astrobiology.

Astrobiology is a new and growing field that not everyone knows about. In general, it is about studying biology in space. This can include studying how to grow plants in space, the effects of spacelike conditions on lifeforms, and trying to find signs of life elsewhere in the universe.

Being an astrobiology freshman in college was when I was introduced to the opportunities in space that would change my life and bring me to where I am today. The first opportunity was to apply for a position on an analog astronaut crew.

An analog astronaut is someone who simulates being an astronaut in a spacelike environment on Earth to prepare for future missions in space. Analog habitats include the Utah desert which resembles the Martian surface and is where the Mars Desert Research Station (MDRS) run by the Mars Society resides. I was the alternate Health and Safety Officer (HSO) for MDRS Crew 239, and I am now the crew commander for Florida Tech's first mission to the University of North Dakota's Inflatable Lunar/Martian Analog Habitat (ILMAH). For two weeks at ILMAH, I will be living as an astronaut, performing EVAs, and creating an educational documentary about our mission.

As an aspiring astronaut, I earned my Open Water SCUBA Diver certification in Summer 2022 and am planning to apply for the PoSSUM Academy for Spring 2023. PoSSUM stands for Polar Suborbital Science in the Upper Mesosphere and is a project aimed towards training the next generation of citizen-scientist astronauts who will be performing research on the state of the upper mesosphere.

The second opportunity was to join the Palmer Lab of Chemical Ecology and Astrobiology, which hosts projects such as CERES and FarmBot. CERES (Controlled Environment Research and Education Space) is a proposal for an isolated module at Florida Tech, open to students and faculty for isolation experiments, immersive education, and analog missions.

FarmBot is a machine that grows crops on its own in simulated Martian dirt, which is known as Martian regolith simulant. Last year, we made the first ever Martian ketchup in collaboration with the Heinz ketchup company by growing tomatoes in Martian regolith simulant. The tomatoes were shown to be safe for consumption and even produced viable seeds for future generations of tomatoes. I've presented my research at the American Society of Plant Biologists conference and the Northrup-Grumman Student Design Showcase.

As the president of ARES, I have presented to and met Mike Moses, the president and CEO of Virgin Galactic, as well as the USRA (Universities Space Research Association), the Ministry of Space Education/Antarexxa Space Global, and college administrators and faculty. As the director of the marketing and outreach program, I plan events and activities for the club, such as our annual Space Week celebration.

I was notified that I was awarded the Astronaut Scholarship while I was doing research on tide pools at the Oregon Institute of Marine Biology (OIMB) in Summer 2022. Later in August, I got to attend the Astronaut Scholar Gala and Conference where I met other Astronaut Scholars, astronauts, scientists, engineers, and mathematicians. I learned about the amazing research other students were doing. As an Astronaut Scholar, I have been able to connect with CEOs, astronauts, and professionals through a professional mentorship program.

As I mentioned earlier, I did fencing for several years. What I did not mention was that I was always the only girl in the class. Well, one time another girl joined the class for a day, but then she immediately quit. For women in male-dominated environments, it can be hard to find the courage to stick to it. The same thing applies to being a woman in STEM. However, as I dig myself further and further into aerospace, the more inspiring women in STEM I come to meet. I was invited to represent Florida Tech at *Reinvented* Magazine's 2022 Space Gala, which is a celebration of women and girls in STEM. At that gala, I met incredible women in aerospace such as Jane Ponyter, one of the original explorers in Biosphere 2 and a co-founder of Space Perspective, and I had never before seen so many women scientists, engineers, and mathematicians all in one place.

I am so fortunate and thankful to be where I am today in the space sector, and I can't wait to see where I end up within the next few years. All of this only happened within the past three years of me being in college, so I can't imagine where I'll find myself in the next three.

Merryl Azriel

"Whether you want a career or a hobby, if the right fit isn't out there, you can create it."

– MERRYL AZRIEL

My first cinema movie was *Star Trek: the Motion Picture.* I was an infant at the time and don't remember it, at least from that viewing, but I like to credit it as the start of my space journey. Growing up, I didn't have a lot of exposure to the real space industry – barring that discarded space shuttle tile of my dad's that I took to show-and-tell whenever possible – but I made up for it with sci-fi books and *Star Trek*.

I was good at English, math, and science and decided on taking a practical path by majoring in engineering with its near-guarantee of good job opportunities. I didn't think I was good enough at physics to try for aerospace engineering (in retrospect, I was totally good enough – and so are you), so I selected low-risk chemical engineering and followed a predictable path from there with a first career in personal care product R&D.

Eight years in, though, I was getting bored and ready for something new so I started reviving my long-buried wish of working in the space industry. How to break into an entirely new career field? I quit my job, moved to France, and enrolled in the space studies master's program at the International Space University. I learned about all the pathways to space careers that I hadn't really seen before, had an amazing astrochemistry internship at NASA, and fulfilled a long-time dream by becoming a magazine editor with Space Safety Magazine. A heady couple years of opportunities later, I found myself working as a communication professional at a NASA contractor. This put me on a path to a US contractor communication and marketing career that drifted away from space. I still enjoy the occasional Venus conference and cheer on missions from the sidelines.

Some secondary and post-secondary schools make it really easy to discover space careers for a wide variety of interests. My schools weren't among them, and I didn't find out until later that I could work in space as an English, finance, biology, geology, arts, or chemical engineering major just as well as an aerospace engineering one (I still get really exciting when I spot an astronaut with a chemical engineering degree!). It is pretty easy to get internships and early career opportunities when you are starting out – take advantage of those options. Even if you end up walking a different path, you will have an amazing experience to look back on and you'll know your options before you make career decisions. It's also never too late to make a change – there are more space organizations and jobs now than ever before. Whether you want a career or a hobby, if the right fit isn't out there, you can create it.

Sara Sabry

"I ... hope to inspire younger generations to dream bigger."

– SARA SABRY

My name is Sara Sabry and I was the first Egyptian astronaut, the first Arab woman in space, and the first woman from the African continent to go to space. I was selected by Space for Humanity for my leadership experience and potential for global impact out of thousands of qualified applicants around the world. I flew to space on Blue Origin's sixth human spaceflight on August 4th, 2022 on their New Shepard rocket (NS-22). The goal of the mission was to experience the Overview Effect and bring this new perspective to benefit life on Earth.

I am the founder and executive director of Deep Space Initiative (DSI), a non-profit company that aims to increase accessibility and opportunity in the space field while enabling deep-space exploration for all humankind. DSI's mission is to help erase the borders on the map by providing opportunities in research and education that are open to all audiences, no matter their nationalities or resources.

I am currently pursuing my PhD in Aerospace Sciences at the University of North Dakota and working at their NASA-funded Human Spaceflight lab. My research focuses on engineering the next generation of planetary spacesuits, and I have aspirations of obtaining my private pilot license along the way. I became Egypt's first female Analog Astronaut in 2021 after completing a two-week moon mission simulation as the crew medical officer at Lunares in Poland and in 2022, I successfully completed the IIAS astronaut training program with Project PoSSUM – a suborbital research qualification program – making me the first female Egyptian trained to do research in space. I received my mechanical engineering bachelor's degree in 2016 from the American University in Cairo, with a concentration in Mechatronics as well as minors in biology, chemistry, and pre-med. I received my master's degree in biomedical engineering in 2020 from Politecnico Di Milano in Italy. My wide range of experiences span from mechatronics and robotic surgery to stem cell development and bioastronautics. After my studies, I worked at a tech startup in Berlin as the deputy CTO and solutions engineering, focusing on augmented reality mobile applications.

I am passionate about increasing the representation of Egyptians, especially women, in the space field and hope to inspire younger generations to dream bigger. As part of my efforts, I co-founded the Space Ambassador Program with the Egyptian Space Agency (EgSA), and I am the first Astronaut Health & Performance Space Ambassador. I continue to work with EgSA on a number of exciting projects, including building the first analog research station in Africa. I was the team lead for the transportation on the moon group within the Lunar Commerce & Economics of the Moon Village Association (MVA), a PoliSpace external advisor, an independent researcher with Biofrequency Analytics, and a Human Factors teaching associate at Mars University.

Prior to my work in the space industry, I was a yoga instructor and CrossFit coach during my undergraduate studies and spent some time practicing mixed martial arts. In 2016 I traveled around Africa, volunteering in different schools and organizations, including a women empowerment center in Uganda and a primary school in South Africa. As a firm believer in protecting our oceans, I am an advanced open-water-certified diver and spent six weeks in Madagascar as a marine conservationist. Check out my website for additional information about my story.

astrosarasabry.com

Nehal Gajjar

"Believe in yourself, embrace your passions, and never be afraid to reach for the stars."

– NEHAL GAJJAR

I grew up in India, captivated by the airplanes flying overhead near the local airport. Like many kids, I wanted to be a pilot. But as I grew, I realized that my curiosity and interests went far beyond just flying planes. I wanted to understand how they worked, how to make them better. That realization shifted my focus to aerospace technology, and I knew I had to go where the opportunities were – the United States.

Buoyed by support from exemplary teachers to mentors and with a mix of excitement and nerves, I left India to chase my dream, not fully knowing the path ahead.

The transition wasn't easy. Moving to a new country, adapting to a different culture was initially tough. But I pored over my studies, first at Georgia Tech and then at Carnegie Mellon's Tepper School of Business. I dived into jet engine technology, spending countless late nights studying the complexities of thermal, mechanical, and vibrational loads. There were definitely moments of self-doubt. I'd wonder if I was smart enough, if I belonged when surrounded by industry experts. But I just kept going, pushing through because I knew I couldn't let fear stop me.

Along the way, I had mentors who believed in me, even when I wasn't sure of myself. One of them is Tom Nangle, former President of Pratt

& Whitney Technologies. To this day, he mentors me. The mentors' support was crucial, and it helped me see that my curiosity didn't need to be limited to aerospace. I started leading research teams exploring how technology could be applied in undersea environments, expanding my horizons even further.

But entrepreneurship was where things really got real. Starting multiple companies wasn't glamorous – it was gritty. There were failures, financial stresses, sleepless nights, and plenty of moments where I wondered if I was making the right choices. But through all that, I learned how to lead, how to take risks, and, most importantly, how to bounce back from setbacks.

In 2019, I stumbled onto something big with iMetalX. We conceptualized a platform to simulate and evaluate for satellite communication in a mesh topology, pushing the boundaries for space applications.

Entering the Hyperspace Challenge was a turning point for me and the company. We were inspired to serve the Space Force. We didn't just gain exposure; we worked alongside brilliant minds at the Air Force Research Lab, refined our roadmap, and identified real market needs. It was a game-changer that opened doors to contracts with federal agencies and accelerated our growth.

But iMetalX is more than just a company to me. It's a way to make space safer, to reduce the constant need for satellite launches, and to protect astronauts. My vision is to make space more accessible for everyone, and I truly believe that technology can help us achieve that.

Now, with over two decades of experience across aerospace, renewables, power generation, and software industries, I've learned to enjoy driving growth and innovation. I've founded multiple profitable R&D startups, secured patents, and re-engineered processes to speed up time to market for aerospace components. But none of this came without struggle. It's been a journey of learning, making mistakes, and finding the grit to keep moving forward.

I've also had the privilege of mentoring other startups through Carnegie Mellon's Schwartz Center. It's rewarding to give back and help guide the next generation of entrepreneurs, especially because I know firsthand how challenging it can be.

Looking forward, I'm excited about the opportunities in space. With SpaceWERX Orbital Prime, iMetalX is exploring new ways to service, assemble, and manufacture in space. The potential for growth in the space economy is huge, and I'm eager to play a role in shaping it. Sustainability is a huge focus for me – I want to see us integrate it into every aspect of space exploration, from design to recycling.

At the end of the day, my story is about more than just professional success. It's about following your passions, no matter how challenging the path may be. I've always believed in leading a balanced life, and outside of work, I love being active – whether it's basketball, sailing, snowboarding, hiking, martial arts or mountain biking. These things fuel my sense of adventure and keep me grounded.

To all the young girls out there, my message is simple: Believe in yourself. Don't be afraid to go after what excites you. The road won't always be easy, but every challenge is an opportunity to grow. If you stay determined, there's no limit to what you can achieve, whether it's in STEM or anywhere else. I'm proof that with enough persistence, you can reach heights you thought were not possible.

Laura Morelli

> **The important thing is to always know that you're doing your best for what you want, even if there are times in which it seems as if there are no results, it will all work out in the end.**

— LAURA MORELLI

Hi there, nice to meet you all! This is Laura, I am twenty-four (by the time you read it maybe not anymore) and I am here to tell you my story, but I would love to hear yours as well so feel free to contact me!

So here we go.

When I was a kid, I used to dream about many different careers, including becoming an astronaut or an intelligence officer working for some kind of institution against organized crime. Growing up, I slowly realized that not everything was possible, some of these options were incompatible and some others I didn't like as much as I thought I did.

What I did like very much, beyond sports, was hands-on work and scientific subjects. By the time I was thirteen, I had decided to pursue a technological and scientific high school in Italy, which should have helped me to be part of the Italian Military Academy, where I wished to specialize in the Air Force Division. However, what I perceived at that point as a massive disappointment suddenly changed everything: I had to put my glasses on! It was a terrible disaster for me because I could not become a fighter pilot, hence no purpose for me in joining a military academy in the first place. And so I didn't. The whole point of it was to first learn to fly fighter jets and then specialize in space, becoming an astronaut. However, turns out it was not the path I had to follow.

Regardless, I started a scientific high school in Italy and finished my courses in a British school in Barcelona. Here, I had to choose some subjects that I wanted to attend at an advanced level during my last years, and it was quite a hard choice for me because I always felt like I wanted to do everything. I started falling in love with different kinds of literature and learnt to appreciate a great variety of topics.

In the end, I chose to do a mix of scientific and social subjects, including literature, mathematics, physics, etc. In my second-to-last year of high school, I started to wonder what academic path I should follow at university. I liked medicine, economics, engineering, law, architecture, physics, art, etc. I didn't know what to do, like many other people, especially because what I really wanted to do – which was a bit of everything – didn't exist at university. So, if you are in this situation, it is okay. There is no right or wrong decision and positive experiences will come out of whatever decision you will make.

Anyway, I started narrowing down my options, basically by elimination, and I was left with law and engineering. Why couldn't I do both? Why would nobody teach a double degree in law and engineering? This is something that I did not and still do not entirely understand. It would have probably been the best choice ever for me, at least from an academic perspective. I started speaking to people that had studied either law or engineering to get their view on these degrees, and I came to the conclusion that I was probably going to fit more into a law degree. Coming from a British school, I was taking for granted that I was going to study at a British university, and I was in the process of my interviews with top schools from the UK when Brexit happened, and the UK left the European Union. That was another massive disappointment. What would I do now? There was a lot of paperwork to study outside of the Union, a lot of trouble if I needed medical attention for whatever reason, and many constraints to non-UK citizens, etc. Maybe the UK was not the best option for me at that point. And so I didn't go.

I re-thought my options for a second and I realized that in Spain, which was an extremely vanguardist element, it was possible to study different degrees at once. Although law and engineering was not available, nor easy to plan according to timetables, I decided to study two different bachelors: law and international relations. Nonetheless, I did not want to leave my scientific self to the side, and I pursued technical and engineering courses online just for personal curiosity. There was a part of me that did not feel complete or represented enough and had to somehow fill it up with knowledge. But what was I going to do about it? How could I merge law and something more scientific like space? It seemed impossible. However, during my last two years at university I had to work on my two theses, one for law and one for international relations. I decided that I was going to try my best to include scientific topics in both of them.

For one of them, I worked on drones from a political perspective. For the other, I studied the legal implications of space programs in the European Union. These were probably some of the most exciting academic moments that I experienced. So what next? How do I transform it into a professional career? Well, that was not possible quite yet. I did an online course in Business Administration while finishing the theses and enrolled in the Master's of Space Studies in Strasbourg because I believed that with a much larger understanding of all aspects of the space field, I would have been much more appealing as a young professional. It turned out to be true, but I must say that beyond being an investment for future opportunities, it was a great academic choice as well. What better than knowing the space field from all sorts of disciplines? I learnt to get out of my comfort zone, learn about engineering, build a satellite model and a robot that would pick up objects on its own. I was finally able to merge my hands-on capabilities with legal studies and much more, it all started to slowly make sense!

Towards the end of the year, I started looking for professional opportunities, be they internships or full-time jobs. I was extremely lucky and proud to have received offers from several space agencies. I decided to start my "adult life" with an internship at the European Space Agency in an extremely interesting field: space strategy. I recently started and it is merging everything that I have tried to build in my past: it has legal aspects, economics, political and technical considerations as well.

Did I make the right choices all the way? I do not know; I really don't think there are right choices. They all feel somewhat wrong, or incomplete, especially being a very ambitious person and aiming to do them all, learn from all and improve by living all possible experiences. However, in the end, they make sense and help you grow into your future self. I am now wondering what will happen next, once this internship is over. I still don't know, but whatever it is, it's going to be great and I will learn, grow, and contribute a lot!

Sometimes not having a plan many months in advance can be destabilizing, but life turns out to be awesome, independently of the plan that you make in your head, which will most likely change as you move ahead. Life goes as it wants to, and that is beautiful. The important thing is to always know that you're doing your best for what you want, even if there are times in which it seems as if there are no results, it will all work out in the end. You can want to be an astronaut, a space lawyer/strategist like me, a scientist, a politician, an economist, a doctor for astronauts, or you can be an artist and paint, draw, or sing about space.

There are plenty of options out there, and many more that I have not mentioned! If you already know which one you like the best, that's great, but if you don't yet, that's fine. Like they said in *Harry Potter*, sometimes the wand chooses the wizard.

Whether you choose your path, or your path chooses you, enjoy the ride.

Georgia Wuorio

> **"I hope you take my story and realize that even with humble beginnings, while navigating the world in a myriad of difficulties, there is a place for you in aviation and aerospace."**
>
> – GEORGIA WUORIO

was 13 when I turned to my mother and said, "I want to fly airplanes." I can't exactly pinpoint where my love of aviation came from. Especially considering I'd never even ridden on an airplane before. My mother just stared at me and said, "Well, why don't you wait until after we take our family vacation to California, and see if you even like to fly." That was 1991 and, needless to say, I fell in love with everything to do with aviation.

I signed up for a Boy Scout Explorer group (which at that time was the only way a girl could be involved with the Boy Scouts) and learned the ground school information and also took some familiarization flights. Eventually I took some lessons, but as a busy highschooler I let it slip aside as other things took precedence. I never did get my pilot's license but I also never fell out of love with aviation.

It was second grade that I was watching the Challenger fly the first teacher into space and witnessed the disaster live on television. I knew it was bad and that people died, but I don't think a second grader can really comprehend the impacts of something like that.

After high school, I was more focused on getting out of my home than anything else. I worked as a manager in a fast food establishment, then as a customer service representative in a few call centers, followed by what I thought was going to be my career as an insurance underwriter.

It was during my tenure with the insurance company that the terror attacks of 9/11 happened. I was in my early 20's and this time I did understand the impacts of the events. A couple years later, we lost Columbia.

When we fast forward to 2005, you'll find me attending a Jacksonville Jaguars game. I remember standing in the stadium for the opening show. There were Hometown Hero's on the field. All sorts of police, medical, fire, and military standing on the field. I looked around the professional football stadium as Lee Greenwood's "I'm Proud to be an American" played over the loudspeaker. I thought, "I'm here in this stadium, without a care in the world, not even worried about my safety, because of people like that on the field. Because of people like my father who retired from the Navy, my siblings who were either in the military or married to military, my friends who were mostly military or law enforcement. What am I doing to contribute?" I went to see a few recruiters the next day.

Honestly, I only went to see the Air Force recruiter because I didn't want to live on a ship and I'm not cut out to be a Marine or to be in the Army. I was laying the foundation at my underwriting job, so I decided on the reserve. There were a few jobs available at the base I wanted to drill at. I chose among a medical technician, air transportation, security forces, aviation crew chief, and aviation avionics. After a short discussion with my little brother, I chose avionics. This was the decision that changed my life.

I'd never fallen out of love with aviation. All through my teen years and young adulthood, I would visit whatever aviation museum I was near and I would attend airshows when time and money allowed. This choice to do avionics fed my love of aviation in a way I never knew was possible.

Being in the reserve is complicated. You are expected to perform your primary job at home like that's all you do. Then you take time away from your full-time career and your family for training and mission requirements. Though it was sometimes difficult to juggle, I wouldn't have changed a thing.

The Air Force taught me the inner-workings of airplanes. I learned not only how they work but also how they think. In my years working the flight line at Charleston Air Force Base in South Carolina, I remember times when I'd sit on the steps overlooking the ramp filled with C-17 Globemaster III's and watching them take off onto their missions and just thinking, "Darn! This is my job! I get to bring these airplanes to life. I get to see them fly away and then come home. Not many people get to do what I do." And, I'd sit there and just take it in.

I have been truly blessed with that feeling also while overlooking the airfield at McConnell Air Force Base in Kansas and Patrick Space Force Base in Florida. The Air Force not only taught me the intricacies of how airplanes work, it also took me to countries in Europe and Asia as well as various states across the United States. My story doesn't quite end there.

In my transition from Charleston to McConnell, I begrudgingly moved from technician to quality assurance. I didn't want to stop working on the jets. But that's where the opportunity was so I took it. This was the second life-changing decision in my aviation and aerospace career.

I worked in quality for about two years then begged my way back into a technician job. A couple years later, and opportunity presented to move to Warner Robins, Georgia and work for Air Force Reserve Command as a Quality Assurance Functional Manager. This position expanded my technical knowledge into the realm of quality and what that means to safe and reliable flight.

I spent just over three years leading quality teams across the Command. During this time I was a single mother and I completed my Bachelor's of Science Degree in Aeronautics with a Minor in Safety. With the completion of my tour at headquarters, I moved to Quality Assurance Superintendent at Patrick. The maintenance unit at Patrick gave me the opportunity to expand my ability to lead teams and to implement good quality policy. This leads me to my third aviation and aerospace life-changing decision.

I applied as a Quality Assurance Specialist for the National Aeronautics and Space Administration (NASA). NASA is where I have really found my passion, my career, my home. I am able to use all the knowledge and skills I've learned as an aviation technician, quality assurance inspector, functional manager, and quality superintendent. I work closely with engineers to ensure procedures are safe not only for people but also for the hardware itself.

From time to time I work with experimental scientists who are exploring the next frontiers of space technology. But most importantly, I get to see and interact with some of the coolest things going to and coming from space! Since joining NASA, I've been promoted twice in three years. I am learning from great minds who worked the Shuttle program as well as the International Space Station program. I am applying my technical knowledge to Artemis (including Orion and Space Launch System), International Space Station, and Lunar experimental payloads. It is because of my journey through various, and perhaps unconventional, experiences that I have gotten to this point.

And even though it started with a 13 year old's proclamation of wanting to fly airplanes, and it took me many course corrections to get here, I have made it to the Golden Ticket of aerospace careers! I hope you take my story and realize that even with humble beginnings, while navigating the world in a myriad of difficulties, there is a place for you in aviation and aerospace.

Martina Dimoska

66 When an underdog wins, they win for everybody, because somebody has to come through that door and break it open and make it possible for other generations that follow. 99

– MARTINA DIMOSKA

A Dream Beyond the Balkans

Introduction

In the small quiet town of Kicevo, nestled in the heart of N. Macedonia, a young woman named Martina Dimoska had big dreams that reached far beyond the borders of her country or society's idea of what girls can aspire to become. Ever since she was little, Martina always knew she had this inner calling, almost like she was destined for big things, as she pioneered multifaceted ideas, bringing diversity along the way.

On the path to her ultimate pursuit of truth, she stumbled upon the vastness of space, which contained many intriguing questions ready to be answered.

But growing up in a region where there was never a space agency or a notable space industry, no Space Heroines and female representation, Martina soon understood that there are almost no opportunities for girls like her to pursue their dreams. But she was determined, as she set out to pursue her dream on a labyrinthine route faced with many challenges.

Dreaming Big in a Small Town

Martina's journey to the stars wasn't easy. In the Balkans, where she grew up, many people believed that space was something far-fetched, unreachable, and ultimately something that only guys can aspire to perhaps achieve, sometime in the future, 500 years from now. But deep down in her heart, she knew she wanted to be a part of something bigger, something that would change the world, and she knew she was cut for it.

Challenges Along the Way

For Martina, finding a way to pursue her dream meant grabbing every opportunity that came her way, no matter how scarce or challenging it seemed. There were no straight paths or clear directions. Instead, she navigated a chaotic path, always searching for ways to learn more about space and science.

In a world where traditional gender roles often limited what girls could do, Martina was determined to break through those barriers. She wanted to show that girls are allowed to be multifaceted, they could could be communicators, scientists, engineers, leaders, and adventurers – all at the same time. But she knew she had to work harder and fight against the odds to prove herself.

Success isn't a straight line

Success

Success

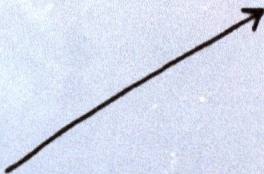

what people
think it looks like

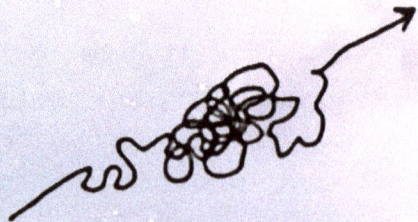

what it really
looks like

Breaking Barriers and Embracing Opportunities

Despite the challenges, Martina's determination never wavered. She worked hard in school, and her dedication paid off when she earned a full 'A-type' scholarship awarded by the Ministry of Education and Science in N.Macedonia to study Material Engineering and Nanotechnology at the Faculty of Technology and Metallurgy in Skopje. If she didn't secure the high GPA Scholarship since the very first year, she was never going to have an opportunity to attend college. This was just the beginning of her streneous journey towards a Space Exploration Path. Miss Martina became a GLOBAL UGRAD Alumna, an exchange student at Kent State University in United States, attending the Aerospace Campus. Afterwards Miss Martina got a rare scholarship from Aldrin Family Foundation, established by the Second Moon Walker, the legendary Buzz Aldrin, to attend Commercial Space Studies Graduate Program at Florida Institute of Technology. Later on Miss Martina received a tuition scholarship from International Space University (ISU), attending the renowned Masters of Space Studies Graduate Program. Some od her ISU research projects included building a 3D solar sinter and ESA-funded "Starship - Impact on the Satcom Industry."

Dreaming Beyond the Stars

In a short amount of time, Martina Dimoska became an awarded published research scholar, a space engineer, and a pioneering figure in space exploration nationally, regionally, and internationally. She worked on developing cutting edge novel space technologies and conveyed notable research encapsulating material engineering and nanotechnology, novel materials in space exploration, additive manufacturing, in-situ resource utilization (ISRU), Lunar and Martian Regoliths and Lunar and Martian Regolith Simulants.

One of her most admirable historic achievements that has earned her place in history books is that she became the first Macedonian and female Balkan Analog Astronaut, serving on three missions followed by a lot of human spaceflight training as well as experiments for which the team research was the baseline for projects flown on the International Space Station (ISS). She was the commander on Mission M08 SELENE via Lunares, she was serving on ESA-funded ICEE.Space and Astroland Interplanetary Agency Mission called APICES and NASA and CSA supported Mission UND ILMAH 14. Martina got her NASTAR Wings after a successful NASTAR Suborbital Space Flight Certification as previously trained Military Pilots, Astronauts and Private Astronauts from the likes of SpaceX, Axiom Space and Virgin Galactic.

Martina knew how difficult the path towards space exploration is when you're coming from a region that has no acces to it, and in that spirit she founded the International Space Alliance (ISA), an organization breaching the gap between agency, industry, and academia - bringing space opportunities to developing regions, and accelerating equal access to Space Exploration. She became the President of ISA with a devoted mission to make sure space is truly accessible for all, especially for those coming from underrveloped regions, scarce resources and underprivileged communities.

Martina's work got her a lot of recognition as well as prestigious titles such as becoming the first Google Woman Techmaker Ambassador in her country, a 3-times TEDx speaker, and she is a recipient of numerous awards, including Forbes 30 Under 30 Nomination for Europe, Top 50 Bloomberg Adria 2023, and Woman or Scientist of the Year and a Person of the Year a plethora of times, awarded throughout various organizations. Amongst her many notable space awards and nominations, she received the distinguished Emerging Space Leader from the International Astronautical Federation where she is also currently involved throughout several committees.

Martina has been recognized internationally by Forbes, Bloomberg Adria, Huffington Post, UNDP, UN, UN Women, OSCE (Organization for Security and Co-operation in Europe), Pineapple Empire, Al Jaizira, TED and TEDx, and many notable others.

Her work is globally recognized, from publications on space exploration to advocacy for women in STEM. As a STEM educator, Martina promotes diversity and inclusivity in space exploration through workshops and social media outreach via her popular handle across social media.

@astro_smarta

Martina's story is not just about reaching for the stars – it's about showing that anyone, no matter where they come from or what challenges they face, can achieve their dreams. Although the constant underdog, through her hard work, perseverance, and a belief in herself, Martina Dimoska has become a trailblazer in space exploration. Her journey inspires us all to dream big, break barriers, and reach for the stars. When an underdog wins, they win for everybody, because somebody has to come through that door and break it open and make it possible for other generations that follow.

About Gary L. Gilbert

Gary L. Gilbert has a passion for space and the Space Industry. He was inspired to compile these stories of women working in roles in that industry which are NOT as an astronaut.

Gary trained at the International Space University in France. There, he saw the struggles of young women who wished to enter the industry but were held back by lack of funds to pay for the necessary study.

His dream was to inspire tween and teenage girls to reach for the stars and have a fulfilling career in the Space Industry, even if they didn't qualify for the Astronaut program. There are so many more positions available outside of that for young people if they don't give up on their dreams!

Gary lives in Florida, USA and enjoys the sunny lifestyle there.

www.ingramcontent.com/pod-product-compliance
Lightning Source LLC
Chambersburg PA
CBHW040855210326
41597CB00029B/4850